AUTOMOTIVE CYBER GOVERNANCE

Opportunities & Challenges for Cybersecurity in Connected & Autonomous Vehicles (CAVs)

AJ KHAN

CISSP, CCSK, PCIP,

MBA (Innovation & Entrepreneurship)

Dedicated to my wife

Halah

Who has always been

My Anchor in Stormy Seas

&

enabled Smooth Sailing in Calm Waters

Contents

Table of Figures

Table of Tables

Foreword

In the fall of 2018, I met AJ at the offices of the APMA. By the end of the meeting, we both understood the gigantic task to help the automotive industry embrace cybersecurity. AJ became a mentor in all things cyber-related, and as all good stories go, the rest is history!

These past few years have been a whirlwind; what AJ has managed to accomplish in such a short time is a credit to him as a person. Chairing the Automotive Parts Manufacturers' Associations first-ever Cybersecurity committee, followed by the launch of the world's first Cyber Awards, as the world tackled this global pandemic, and this past year, the launch of the Global Syndicate for Mobility Cybersecurity; bringing together like-minded nations to further the conversation of cybersecurity in all sectors of mobility.

Driven to innovate and succeed are integral to him; it's part of his cyber DNA! With over twenty years of cybersecurity experience and multiple leadership recognitions and awards, AJ is more than qualified to share his thoughts on all-things-cyber. His latest book, *AUTOMOTIVE CYBER GOVERNANCE - Opportunities & Challenges for Cybersecurity in Connected & Autonomous Vehicles (CAVs)* is a valuable tool for all interested in learning about the impact of cybersecurity on the COMPLETE automotive industry. The book is divided into three sections, starting with the journey of a cyber specialists into the automotive sector, then a detailed explanation of cyber 101, finishing off with a crescendo of cyber governace solutions for; supply chain, operational technology and how we, the humans are the weakest link in the complete equation. The book is a refreshing read in the cybersecurity world, and AJ's broad focus is on strengthening the defences and not just the end product; the vehicle.

Life is about making a difference — no excuses, just do it. Sometimes in life, you meet individuals that you connect with both intellectually and spiritually. When that happens, typically...magic occurs. AJ, I am honoured that you asked me to share a few words before the actual heavy lifting begins my friend. Our meeting in 2018 has been a pivotal moment in my career, and I look forward to my continued learning from you, brother.

Colin Singh Dhillon

Chief Technical Officer, APMA & Author of The Three House.

About the Author

AJ Khan, *CISSP, CCSK, PCIP, MBA (Innovation & Entrepreneurship)*

AJ Khan is the President of the Global Syndicate for Mobility Cybersecurity (GSMC) and the CEO of Vehiqilla Inc., an automotive cybersecurity startup. He is also the former Co-Chair of the Cyber Security Committee (CSC) of APMA. He is a frequent contributor to CAV discussions with entities such as Transport Canada, APMA, GSMC and the European Union Information Security Agency (ENISA).

AJ has over 20 years of experience in Governance, Risk & Compliance, and 13 years of experience in Cybersecurity Innovation and Emerging Technologies. During this time, AJ has worked on Cybersecurity & GRC in Middle East and North America in multiple sectors including Financial Institutions, Oil & Gas, Telecom, Government, Retail & Manufacturing sectors. AJ's Cyber expertise includes building Cyber Governance strategies, Standards & Frameworks, Payment Card security, Cloud security, Industry 4.0, Cyber Supply Chain Risk Management (C-SCRM) and Automotive Cybersecurity.

AJ is ardent advocate of Cyber Security and Governance, Risk and Compliance (GRC). In his view, there is a critical need to address Cyber Security in today's Connected and Cloud deployments and if this effort is not made to address this issue, corporations will incorporate considerable Security Debt that will manifest itself later in the form of major security breaches. This will not only lead to loss of critical data and disruptions in operations activity, it will also lead to a major loss in an organization's brand equity and reputation. Hence, Governance, Risk & Compliance is imperative for success in today's technology and business eco-systems.

Acknowledgements

I can trace the beginning of this book to a single afternoon in the fall of 2018 when I walked into the offices of the Automotive Parts Manufacturers Association (APMA). Both Flavio Volpe, the President of APMA, and Colin Dhillon, the Chief Technology Officer (CTO) of APMA, have been instrumental in enabling my journey on the path of automotive cybersecurity. Their vision, foresight and passion to enable the Canadian automotive sector to thrive in the brave new world of Connected & Autonomous Vehicles (CAVs) has created entirely new sub-sectors of the Mobility Industry in Canada. Indeed, with the Project Arrow initiatives, Canada's first Zero-Emmission Electric Vehicle (EV), APMA has created the supply chain of the future for the Canadian automotive sector.

I would also like to thank my team for helping me complete this book. Raneez Ahmed and Deepan Dhingra gave me valuable technical feedback to enhance the content of this book. Omi Haque went through the entire book with a magnifyinhg glass and dotted the I's and crossed the T's. Ateeq ur Rahman has always been my go-to person for graphic design and he took on the ownership of all the illustrations in this book. All the team worked hard to ensure that the reader finds this book meaningful and I thank them for their tireless effort.

I would also like to thank my wife, Halah, who has always supported my endeavors. During the pandemic, my "author" time was early mornings or late in the evenings, and Halah always enabled this effort in her own unique manner.

Finally, I want to thank the reader. I hope you find this book insightful and helpful in your own automotive cybersecurity journey.

AJ Khan

February 2022

Section 1: The Journey Begins

Chapter 1: Why Automotive Cybersecurity

Chapter 2: The Changing Nature of Mobility

Chapter 1: Why Automotive Cybersecurity?

Chapter Overview

The goal of this chapter is to provide the background of my journey into the Automotive Industry and the path I took to recognize the challenges faced by the Automotive sector from lack of Cybersecurity. I also introduce the APMA (Automotive Parts Manufacturing Association) & the APMA CSC (Cyber Security Committee). We go through the CSC's main objectives and how it has played a critical role in enabling the awareness of cybersecurity at a global level. Also, I detail the creation of Vehiqilla Inc. to address the challenges in the operations side of the Automotive sector. I have also included the importance of the Automotive sector in the current world economy and how much it can impact global prosperity. Finally, I have enumerated the readers who can benefit from reading this book.

Background

Although, I had been exploring cybersecurity relating to Connected & Autonomous Vehicles (CAVs) for some time, I got a truly remarkable challenge and a great learning opportunity to explore this area as the Co-Chair of the Cyber Security Committee (CSC) of the APMA[1] (Automotive Parts Manufacturing Association) Canada. I was someone whose only interaction with a vehicle had been to buy an awesome car and then enjoy the drive in it. I had no previous detailed knowledge of the components and the processes that made such an amazing machine work. Also, my main exposure to the engine of any car that I had owned was to open the hood to add wiper fluid to it. Thus, my knowledge of the inner workings of vehicles was minimal.   .

However, as a Cybersecurity professional for the last 20 years, I had always understood the critical need to fully comprehend any eco-system that required protection from cyber-attacks and other cyber incidents. This eco-system might be a single computer, a corporate network, an entire Data Center, a Payment

---

[1] https://apma.ca/

Cardholder Data Environment (CDE), or in this case, an eco-system built around a connected vehicle. This led me on a journey to understand the modern vehicle i.e., the Connected Vehicle (CV) and the soon to be ubiquitous, the Autonomous Vehicle (AV), also known together as the Connected & Autonomous Vehicles (CAV). My book is the result of this journey.

APMA CSC

I first became familiar with the term Automotive Cybersecurity in mid-2018 when I started interacting with Colin Dhillon at the APMA. As the Chief Technology Officer of APMA, Colin was also leading APMATEC, the Auto-tech arm of APMA. Our discussions centered upon the lack of awareness and the need to advance Cybersecurity in the Automotive sector in Canada. We discussed various ways of addressing this challenge and our discussions led to the formation of an APMA Cyber Security Committee (CSC) at the beginning of 2019. This committee was Co-Chaired by Colin and myself. It's core objective being to make Canadian automotive companies competitive in a globally changing environment by enhancing the awareness of Cyberaecurity in the Canadian Automotive sector and providing guidance on enabling Cybersecurity in the automotive organization.

Below is the original mission of the Cyber Security Committee (CSC) of the Automotive Parts Manufacturing Association (APMA)

"The CSC would assist with providing guidance and best practices to Canadian automotive part suppliers. To help support the safety/security culture by providing best practices throughout individual organizations. Our companies and organizations need to understand how cybersecurity risk affects a company's bottom line and can drive up cost and affect revenue. CSC should provide its expertise in determining the best practice for securing not only the products being manufactured within a factory, but the buildings, its employees and all IoT equipment. The threat is real and our factory floor systems, the engineering offices are all weak links in safeguarding technical/intellectual property information.

We are laying the foundation for a safety/security culture. CSC will look to provide a governance model framework, a scorecard and what could be best described as a 'toolkit'."

Throughout 2019 and 2020, the APMA CSC worked towards enabling a culture of cybersecurity in the automotive sector. A major initiative was the APMA Cyberkit 1.0 which provided a roadmap for automotive organizations to enable cybersecurity. We also had cybersecurity panels in APMA's annual conference in 2019 and then had our own Cybersecurity conference later in that year. Later, Cyberkit 2.0 came out which included modules on ISO 21434 Self-Assessment (Authored by me), V2X, Hardware Security and OT Security. Finally, during the COVID19 pandemic, APMA CSC took on a leadership role and delivered webinars to help the APMA membership to address the new cybersecurity challenges. These webinars covered a range of important cybersecurity topics such as Work-From-Home Cybersecurity, Manufacturing Cybersecurity, and Cloud Security.

Establishment of apmaIAC

The APMA Institute of Automotive Cybersecurity[2] (apmaIAC) was founded in May 2020 during the COVID19 crisis to further the journey of the Automotive Manufacturing towards a Cybersecure mindset. Since we started the journey through the APMA CSC, and during COVID19 Work-From-Home, it became noticeably clear that organizations needed more guidance on enabling Cybersecurity in their environment. The creation of apmaIAC was a direct result of this appreciation.

apmaIAC had four pillars:

- Governance: Enabling Governance Frameworks in the field of Automotive Cybersecurity
- Assessments: Using Assessments against specific Governance Frameworks to highlight and understand the gap
- Education: Furthering the knowledge of Automotive Cybersecurity
- Technology: Enabling new technological solutions to meet the challenges of Automotive Cybersecurity

apmaIAC also launched the Cyber Mobility Awards which were the first global awards to recognize achievements in Automotive Cybersecurity. These were

---

[2] https://apmaiac.ca/

awarded in October 2020 to celebrate the month of Cybersecurity awareness and were greatly acclaimed throughout the Automotive sector.

Establishment of Vehiqilla Inc.

Vehiqilla[3] Inc. was founded by me in June 2020 to address the dearth of companies concentrating on automotive cybersecurity. Although, there are many cybersecurity companies, they are all focused on providing generic cybersecurity services which do not address the specialized needs of cybersecurity in a Connected & Autonomous Vehicle (CAV).

Vehiqilla provides a range of cybersecurity services across the entire sprectrum of cybersecurity requirements for the automotive sector. These include Cyber Governance, Vehicle Incident Management, ECU / APP Security Assessments, V2X Security, Penetration Testing & Vulnerability Assessment. However, at its core, the company operates the Vehiqilla application. One component of this application is the Vehi-SOC module that has been developed to provide real-time cyber monitoring of fleets of CAVs and forms the underlying foundation of the company's Vehicle Security Operations Center (VSOC). Another major component of the Vehiqilla application is the Vehi-Assure Program which focuses on enumerating the cyber risks inherent in the electric and electronic components of the CAV. Extensive research has also been started on the Vehi-Protect range of products which focus on protecting the data-in-flight from the vehicle.

Establishment of GSMC

The journey continues with the establishment of the Global Syndicate for Mobility Cybersecurity[4] (GSMC) in mid 2021. The Global Syndicate for Mobility Cybersecurity (GSMC) is an independent and impartial not-for-profit global organization focused on advancing mobility cybersecurity by bringing together all forms of transportation (of people and goods) through unified security, privacy, and cyber-safety transformation. GSMC will be the centralized global hub, working collectively with multiple jurisdictions on:

---

[3] https://vehiqilla.com/

[4] https://gs4mc.org/

- Global Standards & Regulations
- Public and Private Partnerships (P3s)
- Academic Research and Cyber Innovation
- Cyber Workforce

GSMC SECTORS

Survive & Thrive in a Connected World through Cybersecurity Resilience

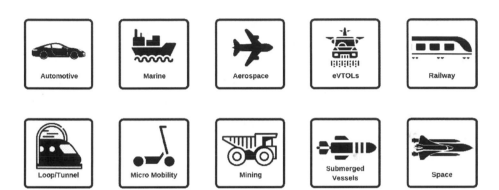

Figure 1 GSMC: Mobility sectors for Cybersecurity Challenges

Industry Recognition

During the last few years, I have been able to work with various industry stakeholders to further automotive cybersecurity. This includes participating in various panels, conferences, podcasts & webinars to enumerate the challenges we face as a society if we do not address automotive cybersecurity. I was invited as a contributor to several workshops arranged by regulatory bodies such as Transport Canada[5] & ENISA[6] (European Union Agency for Information Security). I was truly humbled when all this effort was recognized by Automotive News Canada[7] when this respected auto sector publication named me among its 2020 All-Stars as a Cybersecurity Champion. Another such moment came at the end of 2021 when

---

[5] https://tc.canada.ca/en
[6] https://www.enisa.europa.eu/
[7] https://canada.autonews.com/

APMA awarded me the prestigious Donald S Wood Award for being a Cybersecurity Leader in the automotive sector. I am truly thankful to both APMA and Automotive News Canada for recognizing the importance of Automotive Cybersecurity and ensuring this critical message is heard across the sector.

Aim of this Book

This book aims to educate two different types of audience about Automotive Cybersecurity. The first are those who have been working in the automotive field and for whom Cybersecurity is a new learning curve. This group already understands automotive terminology, especially when it comes to the Connected and Autonomous Vehicles (CAV) but have no knowledge of Cyber concepts. The second group is composed of Cybersecurity Subject Matter Experts (SMEs), like myself, who might have years of Cybersecurity experience but are not aware of the nature of the Automotive Sector and the components that together create the modern vehicle. Thus, the intended audience of this book are all the stakeholders involved in building the "new" Automotive industry and making is cyber resilient.

Some examples of the roles in the Automotive sector who might benefit from insights in this book are

- Chief Executive Officers (CEO)
- Chief Financial Officers (CFO)
- Chief Operating Officers (COO)
- Chief Information Officers (CIO)
- Chief Information Security Officers (CISO)
- Information Technology Managers
- Vice Presidents (Information Technology)
- Vice Presidents (Information Security)
- Information Security Managers
- Cybersecurity Managers
- IT Security Managers
- Research & Development Managers
- Innovation Managers
- Automotive technology Entrepreneurs

- Automotive Industry Association Executives
- Supply Chain Management executives
- Information Security Auditors
- Cybersecurity Architects
- Vehicle Security Architects
- Vehicle Security Manager s
- Fleet Incident Managers

What is the Automotive Industry?

Today, there are almost one and a half billion cars and light trucks around the world. Every year, more than 90 million new vehicles are added to our roads. This leads us to the question of who exactly is involved in enabling this flow of transportation in our society and how do to define the automotive industry in today's world.

The term automotive was created from Greek "autos", and Latin "motivus" to represent any form of self-powered vehicle. This term was proposed by SAE (Society of Automotive Engineering) member Elmer Sperry and first came into use in 1898.

The automotive industry covers a broad range of companies and organisations engaged in this sector. These include organizations involved in the design, development, manufacture, marketing, and selling of motor vehicles, towed vehicles, motorcycles and mopeds. However, the term automotive industry generally does not include industries dedicated to the maintenance of automobiles following delivery to the end-user, such as repair shops and motor fuel filling stations.

It is one of the world's most important economic sectors by revenue and is critical to the economy of many developed countries. As per Behzad Saberi[8] in International Robotics and Automation Journal

---

[8] http://medcraveonline.com/IRATJ/IRATJ-04-00119.pdf

"According to the world association of car manufacturers "OICA" in 2017, 73.4 million cars and 23.84 million trucks were produced in the world. According to international estimates, the average annual turnover of the world automobile industry is more than 2.75 trillion Euro, which corresponds to 3.65% of world GDP. In the automotive industry over the last ten years (2007-2017) there was a 25% increase in production. Cars are one of the world's largest export products, surpassing oil revenues, for example, world car exports by country in 2016 estimated at 698.2 billion US dollars. The industry is also a major innovator, investing more than 84 billion euros in research, development and production."

The above excerpt shows that the automotive sector plays a critical role in the economies of many countries. Indeed, for developed countries, the share[9] of the automobile industry in the GDP ranges from 5 to 10% if consumption of output of related industries is included.

The below table shows the production statistics for 2020 given by OICA[10] (International Organization of Motor Vehicle Manufacturers)

---

[9] http://medcraveonline.com/IRATJ/IRATJ-04-00119.pdf

[10] https://www.oica.net/category/production-statistics/2020-statistics/

Table 1 OICA Vehicle 2020 Production Statistics

Country	Cars	Commercial Vehicles	Total	% Change
Argentina	93001	164186	257187	-18%
Austria	104544	-	104544	-42%
Belgium	237057	30403	267460	-6%
Brazil	1608870	405185	2014055	-32%
Canada	327681	1048942	1376623	-28%
China	19994081	5231161	25225242	-2%
Czech Republic	1152901	6250	1159151	-19%
Egypt	23754	-	23754	28%
Finland	86270	-	86270	-25%
France	927718	388653	1316371	-39%
Germany	3515372	227082	3742454	-24%
Hungary	406497	-	406497	-18%
India	2851268	543178	3394446	-25%
Indonesia	551400	139886	691286	-46%
Iran	826210	54787	880997	7%
Italy	451826	325339	777165	-15%
Japan	6960025	1107532	8067557	-17%
Kazakhstan	64790	10041	74831	51%
Malaysia	457755	27431	485186	-15%
Morocco	221299	27131	248430	-38%
Mexico	967479	2209121	3176600	-21%
Poland	278900	172482	451382	-31%
Portugal	211281	52955	264236	-24%
Romania	438107	-	438107	-11%
Russia	1260517	174818	1435335	-17%
Serbia	23272	103	23375	-33%
Slovakia	985000	-	985000	-11%
Slovenia	141714	-	141714	-29%
South Africa	238216	209002	447218	-29%
South Korea	3211706	295068	3506774	-11%
Spain	1800664	467521	2268185	-20%
Taiwan	180967	64648	245615	-2%
Thailand	537633	889441	1427074	-29%
Turkey	855043	442835	1297878	-11%
Ukraine	4202	750	4952	-32%
United Kingdom	920928	66116	987044	-29%
USA	1926795	6895604	8822399	-19%
Uzbekistan	280080	-	280080	3%
Others	709633	109475	819108	
Total	55834456	21787126	77621582	-16%

13

The COVID19 pandemic has had a significant impact on the production of the vehicles the world over. However, as we come out of the pandemic, the automotive sector will continue to be a driver for job creation, job growth, innovation, and economic prosperity for the global economy.

Automotive Industry & this book

This book is meant for all stakeholders of the coming wave of new CAVs. We already have millions of Connected Vehicles on our roads but soon these will become ubiquitous. In addition, Autonomous Vehicles are being tested by many companies and before long, driverless vehicles will start getting deployed in several jurisdictions. Thus, everyone involved in the rollout of these vehicles as well as their operations, maintenance and disposal would need to comprehend the Cyber challenges and requirements to ensure the security & safety of these vehicles. This includes policy makers, regulators, designers, engineers, OEMs, Suppliers, Dealers & Distributors, maintenance workers, fleet owners, Insurance companies, municipalities and indeed, the "driver" of the CAV. Only by ensuring this detailed understanding of the Cyber realities of our Connected Vehicles, will we benefit from this transformation in Mobility.

What you Learned in this Chapter

This chapter was aimed at clarifying the following

- My journey into Automotive Cybersecurity through the creation of APMA CSC, apmalAC, Vehiqilla Inc. and GSMC.
- The aim of this book i.e., to educate the Automotive sector about Cybersecurity and to educate the Cyber SME about the Automotive sector
- The intended readers of this book
- The importance of the Automotive Sector for the growth of the current world economy

Chapter 2: The Changing Nature of Mobility

Chapter Overview

This chapter is aimed at highlighting the historic nature of the transformation happening in the Automotive sector. The automotive industry has come a long way from the birth of this industry in the late 18th century and this is made clear in this chapter by describing the history and the evolution of the earliest automobiles. This chapter then goes on to include the basic automotive terminology used in the traditional automotive ecosystem. It then continues to detail the transformation taking place in the automotive sector and enumerates the new autonomous terminology being utilized in the Connected & Autonomous Vehicles (CAV) eco-system. Finally, this chapter introduces the reader to the emerging area of smartcities and the cybersecurity challenges that will have to be overcome due to this shift in mobility.

History of the Automobile

The earliest "automobiles" were developed in the late 18th century. These were steam-powered, self-propelled vehicles large enough to transport people and cargo. A Frenchman by the name of Nicolas-Joseph Cugnot demonstrated his fardier à vapeur[1] ("steam dray") in 1770 and 1771. This was an experimental steam-driven artillery tractor and proved to have had an unfeasible design. Later, in 1784, William Murdoch, a British Engineer, built a working model of a steam carriage in Redruth and in 1801, another Brit, Richard Trevithick was running a full-sized vehicle on the roads in Camborne. Also, in 1789, the first automobile patent in the United States was granted to Oliver Evans.

---

[1] https://en.wikipedia.org/wiki/History_of_the_automobile#Electric_automobiles

Figure 2 Cugnot's steam wagon, the second (1771) version[2]

During the early 1800s, there were a plenty of innovations to improve the design of the steam-powered vehicles. These included hand brakes, multi-speed transmissions and better steering. Further efforts to develop more usable vehicles were made throughout the century and by 1867, a Canadian jeweller named Henry Seth Taylor had demonstrated his 4-wheeled "steam buggy" at the Stanstead Fair in Stanstead, Quebec. This was succeeded by the development of the first "real" automobile[3] which was produced in 1873 by Frenchman Amédée Bollée in Le Mans. This vehicle was a self-propelled steam road vehicle to transport groups of passengers and was a major step forward in the development of the automotive industry.

---

[2] https://en.wikipedia.org/wiki/History_of_the_automobile#Electric_automobiles
[3] https://en.wikipedia.org/wiki/History_of_the_automobile#Electric_automobiles

Basic Automotive Terminology

In this section, some basic explanations, and definitions of the most fundamental automotive technology are detailed.

Internal Combustion Engine

Although, the early road vehicles used steam-powered vehicles, the Internal Combustion Engine or ICE is used in most automobiles on the road today. An internal combustion engine (ICE) is a heat engine where the combustion of a fuel occurs in combination with an oxidizer (usually air) in a combustion chamber that is an integral part of the working fluid flow circuit and the energy produced enables the vehicle to move forward.

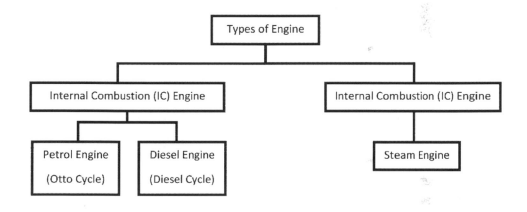

Figure 3 Engines in Vehicles

Some examples of Internal Combustion Engines[4] being used in vehicles today include

- V6: A V-type engine consisting of two banks of three cylinders.
- V8: A V-type engine consisting of two banks of four cylinders, often found in trucks, luxury, and sports cars.

---

[4] https://en.wikipedia.org/wiki/V_engine

- V10: A V10 is an engine with 10 cylinders arranged in two rows of five; some examples include the Audi R8, Lamborghini Sesto Elemento, and the Dodge Viper.
- V12: A V12 is any engine consisting of 12 cylinders arranged in two rows of six cylinders. V12s are largely uncommon and typically found in flagship sedans (BMW 7 Series, Audi A8, Mercedes-Benz S Class) and outlandish exotics like the Enzo, Pagani Zonga, and McLaren.
- V16: A V16 engine is any engine with 16 cylinders arranged in the shape of a 'V' with two banks consisting of eight cylinders each. Modern V16s are rare and have typically been restricted to concept cars and limited run vehicles from smaller boutique manufacturers.

Drivetrain

The drivetrain of a motor vehicle is the group of components that deliver power to the driving wheels. This excludes the engine or motor that generates the power.

Powertrain

The powertrain is considered to include both the engine or motor and the drivetrain.

Transmission

A car's transmission 'transfers' power from the engine or motor to a drive mechanism, often a live axle, through a series of gears and clutch.

Clutch

A device that disconnects the engine from the transmission, which allows the car to change gears and then reconnect again once the gear change has been made.

Manual Transmission

A type of transmission that requires drivers to shift manually from gear to gear.

Automatic Transmission

Type of transmission that automatically changes gears instead of making the driver shift through them manually.

CVT

Abbreviation for Continuously Variable Transmission. This type of transmission does not have "gears" like a manual or automatic.

Cabin

Cabin refers to the interior space of a car.

Dashboard

The area located above and behind a steering wheel that typically houses the instrument cluster and other on-board electronics.

Instrument Cluster

The area behind the steering wheel, often featuring a tachometer and speedometer.

Differential

A set of gears that allows a vehicle's wheels to rotate at different speeds. This is critical when the vehicle is turning as the outermost wheel needs to turn further and faster than the innermost wheel.

Limited Slip Differential

Cars with limited slip differentials can send rotational power to both wheels when one is raised off the ground or experiencing slippage. This is different from Standard or "open" differentials. These cannot perform this task where if one wheel is raised off the ground, it will spin while the wheel in contact with the ground will remain stationary.

Displacement

Displacement refers to the volume of an engine's cylinders and the total air displaced by the pistons inside those cylinders. Displacement generally defines how powerful an engine is and is typically measured in cubic centimeters/liters, and cubic inches. As an example, a "2.8-liter engine" displaces 2.8 cubic liters (or 2,800 cubic centimeters, or "2,800cc") of air in one complete combustion cycle. However, American-made cars measure displacement in cubic inches.

ESC

Short for Electronic Stability Control, ESC monitors a car's traction and provides various counter-measures – like braking and even reducing engine power – to help regain vehicle stability during strenuous driving conditions.

HP

HP is an abbreviation for horsepower, and it refers to the amount of power needed to lift 550 pounds one foot in one second. Horsepower is often used when calculating how fast or powerful a vehicle is but other factors like vehicle weight and aerodynamics also play a role.

MPG

This is the amount (in miles) a vehicle can travel per gallon of gasoline used while an engine is running.

Speedo

Speedo is short for speedometer which is a meter or gauge used to measure how fast your vehicle is traveling. Most speedometers are measured in either miles per hour or kilometers per hour.

Tachometer

Tachometers are typically placed in a vehicle's instrument cluster and measure RPMs (revolutions per minute). This is the number of times an engine's central crankshaft rotates.

Torque

Torque is the pulling power of a vehicle. This is obviously a crucial measure for trucks hauling heavy loads. Sports cars also typically have greater torque as cars with higher levels of torque can also accelerate faster. Torque is measured in pound-feet (lb-ft).

Turbocharging

Turbocharging is the method of increasing air pressure, temperature, and density of the air delivered to an engine (also known as "forced induction"). A turbocharger is powered by a car's exhaust or sometimes by electricity.

Automotive Sector evolution

In the last few years, the automotive sector has evolved considerably. The primary reason for this evolution is the human desire to always stay connected and the catalyst for this evolution are new and exciting technologies that can enable this connectivity. Innovation has also ensured that there are immense synergies in the "Vehicle of the Future" for both the Automotive and Information Technology industries. Examples of these innovations include the environmentally friendly electric vehicles (EVs) and V2X (Vehicle to Everything) connectivity for enabling communications outside the vehicle. Other innovations are happening daily as automotive technology is used to solve new challenges for the Automotive sector. This means that the field of Automotive Technology is projected to grow exponentially for the foreseeable future and ensure that self-driving autonomous vehicles (AVs) develop into more efficient and safer forms of transportation.

Automotive Technology History

Since the Internal Combustion Engine was invented in the 1800's, the automotive industry was perceived to be related to the field of Mechanical Engineering. Building a vehicle was always about steel and plastic. In contrast, the technology industry has always been all about computers & lines of software code and has proliferated into our daily lives from cell phones to ATMs. This proliferation of technology into society has also slowly but steadily entered the automotive sector. Some examples of this technology enablement in vehicles are apps for navigation based on global positioning systems (GPS) and software that enables better design of vehicles. However, there was slow acceptance of this technological evolution from the consumers. Thus, only recently consumers have become accepting of the integration of technology into their everyday vehicle.

Mobility's new frontiers: CASE

Today, we are living through a time when Mobility is undergoing one of the most transformational social, technological, and economic shifts of recent times. This transformation in Mobility is defined as CASE (Connected, Autonomous, Secure, Electric) and is being influenced by disruptive forces unleashed by technologies related to Connected & Autonomous Vehicles (CAVs), Electric Vehicles (EVs) &

alternative powertrains, and on-demand mobility services. New business models and markets are being emerging and whole sectors related to transportation are either converging or even vanishing entirely.

Let us explore each Influencer of CASE in more detail.

Connected

Today we live in a connected world and data is on our fingertips all the time. This ability to "Connect" everywhere has also been applied to enhance our experience in mobility and a number of features have been added to the modern vehicle to make it a "Connected Vehicle" (CV). These include apps for drivers and passengers such as navigation and infotainment apps to more advanced telematics functions. Furthermore, technologies such as ADAS & LiDAR help make the driving experience safer and richer while providing important information in the background to improve the performance of the CV.

Autonomous

An Autonomous Vehicle (AV), also known as a self-driving car, is a vehicle that can guide itself without human interaction. Today, self-driving cars are a reality, and they are being road-tested in many urban test centers. Soon, we will have these "robots" as part of our everyday lives.

There are five levels[5] of Autonomous Vehicles as explained in the image below[6].

---

[5] https://vehiqilla.com/WhitePapers//b47db244-361e-4910-b906-2f4e32eebb79.pdf
[6] Source: Deloitte

23

No Driving Automation	Driver Assistance	Partial Driving Automation	Conditional Driving Automation	High Driving Automation	Full Driving Automation
Level 0	**Level 1**	**Level 2**	**Level 3**	**Level 4**	**Level 5**
There is no automation at this level. All driving tasks are performed by the driver, even when enhanced by active safety solutions.	The vehicle is controlled by the driver, but the vehicle also has a few driving assist features that can provide, for example, steering OR braking/acceleration assistance.	The vehicle has combined automated features that can provide, for example, steering AND braking/acceleration assistance simultaneously, but the driver must remain engaged in the driving task and monitor the environment at all times.	The vehicle has an automated driving system capable of full control of the driving task; however, a driver is still a necessity. The driver must respond to requests to intervene and be ready to take control of the vehicle at all times.	The vehicle is designed to perform all driving functions under certain conditions. At this stage of automation, the driver may control the vehicle in limited conditions.	The vehicle is designed to perform all driving functions under all conditions. It may be optional for a driver to control the vehicle at this stage of automation.
• Blind spot warning • Lane departure warning	• Lane centering OR • Adaptive cruise control (not simultaneously)	• Lane centering AND • Adaptive cruise control (simultaneously)	• Traffic jam chauffeur	• Local driverless taxi • Pedals/steering wheel may or may not be installed	• Same as level 4, but feature can drive everywhere in all conditions

Figure 4 Levels of Autonomous Vehicles

Secure

Although, some organizations use the "S" in CASE for the "Shared" business model of modern vehicles, "Secure" is much more appropriate considering the cybersecurity challenges in CAVs. This is because of the fact that if a CAV is not secure then it cannot be driven as it is a health & safety hazard. However, that realization is only beginning to assimilate into the automotive sector resulting in more emphasis on the "Secure" part of CASE.

Electric

Innovations in power devices, cells. and batteries have made "Electric Vehicles" (EVs) viable for ownership and transportation. Both types of Electric Vehicle, all-electric vehicles (AEVs) and plug-in hybrid electric vehicles (PHEVs) have proved successful commercially and have been accepted by consumers. Also, Electric Vehicle Chargers can now be found easily and this barrier to greater ownership of the EV is slowly becoming obsolete.

CAV Technology Defined

In today's world of Connected and Autonomous Vehicles (CAV), Automotive technology can be defined as the field of integrating technology into self-propelled vehicles or machines. In this brave new world, OEMs, existing suppliers, and innovative start-ups are ensuring that the intersection of automotive transportation and new technologies leads to an enhancement of our "driving" and "transportation" experience. This is a wide-ranging group of technologies and have grown to become major economic eco-systems such as Electric Vehicles and ride-share platforms.

Auto-Tech in Connected Vehicle

This section briefly defines the various technologies used in today's vehicles.

ECU

An ECU or the Electronic Control Unit is used in Connected Vehicles to control the functioning of vehicle components through a computer with internal pre-programmed and programmable computer chips. For example, the vehicle's engine computer ECU is used to operate the engine by using input sensors and output components to control all engine functions.

CANBus

The CANBus is short for Controller Area Network Bus. This is a connectivity protocol that was developed by Robert Bosch and has quickly gained acceptance in the automotive and aerospace industries. The CANbus is basically a serial bus protocol to connect individual systems and sensors as an alternative to conventional multi-wire looms.

VEHICLE WIRING: CANBus NETWORK

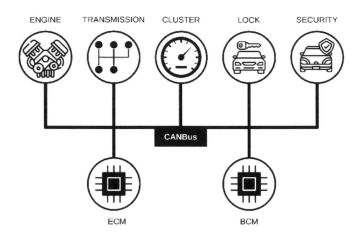

Figure 5 CANBus Network

ABS

Anti-locking braking system (ABS) is an automated safety system that helps a vehicle's wheels continue to rotate during heavy braking. This ensures that the driver does not lose control of the vehicle due to uncontrollable wheel skidding.

ACC (Adaptive Cruise Control)

Adaptive Cruise Control is a form of cruise control that automatically (without any driver input) brakes and accelerates to match the flow of traffic. Some ACC systems can even come to a complete stop and accelerate from a standstill.

LiDAR

LIDAR stands for Light Detection and Ranging and uses laser pulses to build a 3D model of the environment around the car. Essentially, they help autonomous vehicles "see" other objects, like cars, pedestrians, and cyclists, around them.

Lane Departure Warning System (LDWS)

Lane departure systems are a form of onboard safety tech and warn the driver through cameras and onboard sensors whether the vehicle is drifting out of marked lanes. Typically, LDWS involve some type of audio-visual warning for the driver.

In-Car Connectivity

In-car connectivity refers to the technologies that are used when inside the vehicle. Some examples of this are handsfree Bluetooth connectivity for mobile devices, navigation, and the vehicle's multimedia entrainment system.

GPS

GPS stands for Global Positioning System, which is based on a network of satellites in orbit around the earth. The signals from the satellites can be used to pinpoint a location on Earth with remarkably high accuracy. GPS is commonly used in a car's navigation system and in cell phones mapping apps.

NAV System

NAV Systems is an abbreviation for navigation systems and is sometimes also referred to as a GPS system.

Blind spot monitoring

Blind spot monitoring is a technology that scans a vehicle's blind spot for objects. If detected, a warning is given to the driver of the vehicle. This might be just a visual warning or an audible warning depending on whether a turn signal is activated when another car is present in a blind spot.

Bluetooth

Bluetooth is a radio system with the ability to transmit data over short distances, usually a maximum of 30 feet or so. In a car, a Bluetooth connection is typically in the form of wireless music streaming and hands-free phone usage.

Infotainment

A Combination of information and entertainment apps that are provided in the Connected Vehicle. This might include the navigation and stereo interface and employ touchscreens and have limited smartphone app integration as well as social media content such as Twitter and Facebook.

Push button Ignition

Push button ignition (also referred to as push button start) is any system that allows you to start a car by pressing a button instead of physically inserting and turning a key in the ignition.

Key Fob

In the automotive world, a key fob is a small decorative item, often sporting an automaker's logo, that can lock, unlock, open the trunk, and sometimes start a vehicles engine.

Remote Start

Remote start refers to any car that can be started without physically being inside the car and using a key to turn the ignition. An example of this would be to use an app on your smartphone to start the vehicle. However, there are cybersecurity ramifications of such a technology.

Keyless entry

Keyless entry denotes a car that can be unlocked without physically inserting a key to unlock the doors. Typically, this is done using a key fob that enables the system to work.

Smart Key

A smart key is any car key that can lock, unlock, and sometimes start a car's engine remotely.

Touchscreen

A touchscreen is any interactive screen that provides some sort of tactile feedback when pressed. Touchscreens in automobiles are typically used for in-car entertainment and navigation systems.

WI-FI

Cars with built-in Wi-Fi allow a specific number of devices to be connected to the network and for data to be transmitted and received wirelessly to those devices.

LCD Display

A liquid crystal display (LCD display) is a flat panel screen often used to display a car's navigation and entertainment system. Generally, LCD displays contain tiny liquid crystal material between two pieces of polarized glass.

OBD II

OBD II stands for on-board diagnostics and is an industry-standard port found in modern vehicles that is used to connect diagnostic equipment, vehicle data and diagnostic trouble codes through a car's on-board computer. When a vehicle is taken to a maintenance provider, they plug their computer into the ODB II port. App makers can also make use of this port to provide information on the status of various performance metrics of the vehicle.

Regenerative braking

Braking systems typically employed in hybrids and battery electric vehicles that captures energy created during braking and transfers it to the onboard battery.

Semi-autonomous vehicles

Semi–autonomous vehicles refer to vehicles that largely operate on their own and without a normal levels of driver input. These vehicles might have automated features such as adaptive cruise control or lane keeping technology.

Start-Stop

Any car equipped with a start-stop system can automatically shut down its engine when idling and start back up again when the driver takes their foot off the brake. It is often a feature in hybrids and is designed to help increase fuel economy.

ADAS

Advanced driver-assistance systems (ADAS) are electronic systems that help the vehicle driver while driving or during parking.

Telematics

Telematics is a method of monitoring an asset (car, truck, heavy equipment, or even ship) by using GPS and onboard diagnostics to record movements on a computerized map.

Electric Vehicle Terms Defined

Below are some terms commonly used in Electric Vehicles.

EV

EV is an abbreviation for Electric Vehicle.

Hybrid

A hybrid is any vehicle that employs more than one source of power to move the vehicle. An overwhelming number of hybrids on the road today employ a gasoline engine and an electric motor.

Li-ion battery

Li-ion batter is an abbreviation for lithium-ion battery, a type of rechargeable battery used to power electric cars and hybrids.

Level 1 Electric Vehicle Charging Stations

Level 1 refers to electric vehicle charging stations and standard household outlets that deliver 120 volts of AC to an EV's on-board battery.

Level 2 Electric Vehicle Charging Stations

Level 2 chargers refer to public and private charging stations that deliver 240 volts worth of AC to an electric vehicle's on-board battery.

MPGe

MPGe stands for miles per gallon equivalent. MPGe describes the energy efficiency of an EV and plug-in hybrid. The standard conversion used by the EPA is 115,000 British thermal units (BTU) per U.S. gallon of gas, which is equal to 33.7

kilowatt-hours of electricity. Essentially, this measure refers to how far the car could go on a gallon of gas if that was the fuel being used.

SmartCities of the Future

The world is becoming more urbanized, and more than 60% of the world's population is expected to live in cities by 2050. Across the world, cities are using technology to rapidly transform themselves and manage this influx of people to improve all aspects of their citizens' lives. Healthcare, Finance, Education, Logistics & Supply Chain, Entertainment and, indeed, all other facets of human life are being improved through Smart technology initiatives. This rapid trend towards SmartCities is essential for the future of humanity as making the urban centers of the world better places to live in would ensure the sustainability of human life.

The changes in Mobility are an important part of SmartCity initiatives and Infrastructure elements that enable transportation are being deployed in these SmartCities. These technologies are used in automotive transportation but are not necessarily tied to the vehicle itself. Thus, there is a need for Vehicle-to-Everything (V2X) communication to enable smooth functioning of traffic to ensure the lifelines of SmartCities i.e., its transport networks continue providing supplies and transportation to all the citizens of the SmartCity.

Mobility & Cyber Governance Challenges

Data has become essential for smooth operations of all aspects of the SmartCity. This includes the data needed for the efficient functioning of its Mobility component. Privacy, Authentication, Encryption, Cloud Connectivity, Incident response and in-vehicle security challenges for Connected & Autonomous Vehicles (CAV) need to be explored and solutions proposed to ensure that the data used in the CAV is secure and protected. Furthermore, transmission of data during Vehicle-to-Everything (V2X) communication into the SmartCity infrastructure needs to be protected and governed by Cyber Governance frameworks.

What you Learned in this Chapter

This chapter was meant to give you an understanding of the transformation happening in Mobility. This includes the following areas:

- The history of the automotive sector
- Basic automotive terminology traditionally used in the automotive sector
- The transformational changes occurring in Mobility, as defined by CASE
- CAV Technology terminology
- EV terminology
- SmartCities and a brief description about their corresponding cyber challenges.

Section 2: Cyber Challenges for Automotive Manufacturing

Chapter 3: Cybersecurity basics

Chapter 4: Cybersecurity challenges in CAVs

Chapter 5: OT Cybersecurity: Securing Manufacturing Networks

Chapter 3: Cybersecurity Basics

Chapter Overview

This chapter is aimed at the Cybersecurity novice and details the core cybersecurity management concepts starting with the A-I-C Triad. The Cybersecurity novice needs also to understand basic Cybersecurity concepts to apply these to automotive cybersecurity best practices. Privacy is highlighted as a major concern for the automotive sector as tracking of vehicle information could impact the safety of the humans using any vehicle. Hacking also remains a great concern in this sector as Cyber best practices are still not followed in many automotive organizations and leads to incidents of Ransomware and more severe breaches of the organization's data assets.

The A I C Triad

The A-I-C or the C-I-A Triad[1] stands for Availability, Integrity and Confidentiality which are considered the three pillars of Cybersecurity. During a risk assessment process, every threat to a system's Cyber posture is evaluated in terms of its impact to the Availability, Integrity and Confidentiality of that system. In case the risk from a specific threat is deemed above acceptable levels, risk mitigation techniques must be utilized to bring it to a manageable level.

---

[1] https://www.isc2.org/Certifications/CBK#

CONFIDENTIALITY

INTEGRITY   AVAILABALITY

Figure 6 A-I-C Triad

Let us explore each one of these pillars individually.

Availability

In terms of Cybersecurity, Availability is the ability of an eco-system to ensure the reliable and timely access to data or computing resources by the appropriate authorized personnel. The reverse of Availability is Destruction as it makes unavailable the access to data or computing resources to the appropriate personnel as and when needed by them.

There are many aspects to the loss of Availability to a computing resource. This loss could be severe as in the case of loss of data processing capabilities due to natural disasters or human actions or it could be intermittent such as the loss of availability caused by a Denial of Service (DoS) attack. A Denial of Service (DoS) attack happens when an intruder carries out specific actions that tie up computing services and the system becomes unstable or even unusable for authorized users. However, whether the impact to availability is low, medium, or high, an organization needs to have risk-mitigating controls in place to counter scenarios that might have an adverse effect on the Availability tenet of the A-I-C Triad.

These controls can be Physical Controls, Technical Controls, or Administrative Controls.

Physical Controls are designed to prevent loss of availability due to physical damage. These include preventing unauthorized personnel from physically entering restricted areas, ensuring fire & water control mechanisms are implemented, developing Hot, Warm and Cold sites for data processing, and having Off-site backup storage facilities.

Technical controls that address Availability include Fault-tolerance mechanisms, Electronic Vaulting, and Access control software. Electronic Vaulting is the term used for backing up data and transmitting the output electronically to a secured offsite storage location. In the age of the Cloud, this is a common practice and used by most SaaS apps.

Finally, Administrative Controls that can address Availability include Access control policies, Operating procedures, Contingency planning, and User training. All these are best practices for limiting impact to the Availability tenet and can be utilized to prevent destruction of a computing asset.

Integrity

The second pillar of the A-I-C Triad is Integrity. Its aim is to ensure that

- Modifications cannot be made to data by unauthorized personnel or processes
- Unauthorized modifications cannot be made to data by authorized personnel or processes
- The data are internally and externally consistent

The reverse of Integrity is Alteration i.e., modification of data that would adversely impact the Integrity of data. The basic principles to establish Integrity Controls are Need-to-know Access (Least Privilege), Separation of Duties and Rotation of Duties. The principle of Need-to-know Access (Least Privilege) highlights that Users should be granted access only to those files and programs that they absolutely need to perform their assigned job functions. The second principle is that of Separation of Duties which states that no single employee should have control of a transaction from beginning to end and that two or more people should be responsible for performing any transaction. This ensures that no unauthorized modifications can be done to data even by authorized personnel and helps reduce insider threats.

Confidentiality

Confidentiality is the third and final pillar of the A-I-C Triad. Confidentiality protects against the intentional or unintentional unauthorized disclosure of any message and its reverse is Disclosure.

There are many threats to Confidentiality. These include Hackers, Masqueraders, Unauthorized user activity, Unprotected downloaded files, Networks, Trojan Horses and Social Engineering.

A Hacker or Cracker is someone who bypasses a system's access controls by taking advantages of security weaknesses that the system's developers have left in the system. On the other hand, a Masquerader is an authorized, or unauthorized, user of the system who has obtained the identity of another user. Thus, he or she can gain access to files available to an authorized user by pretending to be that authorized user. Another threat to Confidentiality is Unauthorized user activity. This occurs when authorized users gain access to files that they are not authorized to access. In the age of the Cloud, unprotected download of files and data is another major threat to Confidentiality. Downloading data from a secure location to an unsecure system can severely compromise the confidentiality of information. Networks also can be used to compromise the confidentiality of data as data flowing through the networks can be viewed at any node of the network. A good example of this is "Account Harvesting" done on unencrypted user IDs and secret passwords. Another tool that can be used to enable Disclosure are Trojan Horses. They can become resident on a target system and can routinely copy confidential files to unprotected resources which can then be transmitted to a hacker. Finally, Social Engineering is one of the easiest forms of enabling Disclosure as human beings are always the weakest link in any cybersecurity chain. This is a non-technical method of intrusion and relies heavily on human interaction.

Data Classification is important to ensure Confidentiality. Data assets must be classified based on their importance with labels such as public, sensitive, confidential, proprietary, private, and critical. Only when such a classification scheme is in place, data assets can be appropriately protected based upon their classification level and the confidentiality of the data element maintained.

Key Cybersecurity Concepts

Below are a few key Cybersecurity concepts whose understanding is essential for developing a holistic Cybersecurity profile of an organization.

Identity Management

To understand Identity Management, the term of Identification must be clearly defined. For Cybersecurity, identification is defined as the means through which a user's claim to his or her identity to a specific system is verified and validated. Based on this definition, Identity management is a key component of an organization's access control architecture, as it helps validate the identities of the organizations users' before granting them the right level of access to any system and data.

Access Control

Access Control is the discipline of managing a specific user's access to resources to keep systems and data secure and protected. Access Control has three main areas

- Authentication: The testing of the evidence of a user's identity (Identity Management)
- Authorization: The rights & permissions granted to an individual or process after the individual has been authenticated
- Accountability: A system's ability to determine the actions and behavior of a single individual within a system

The terms' identity management, authentication, and access control are used interchangeably. However, it is important to emphasize that each of these functions individually serve as distinct tiers for enterprise security processes.

Least Privilege

As discussed in the Integrity pillar of the A-I-C Triad, another best practice for cybersecurity is the principle of Need-to-know Access (Least Privilege). This principle emphasizes that the access granted to any data element should be limited to the bare minimum access level that is absolutely needed to perform the specific assigned activity. The principle of Least Privilege is applicable to any user, program, or process in a secure environment. Example of this principle are Read, Write & Execute privilege levels that are applied to specific data files for specific user access.

Separation of Duties

Separation of Duties is the second principle of the Integrity pillar of the A-I-C Triad and states that no single employee should have control of a transaction from beginning to end and that two or more people should be responsible for performing any transaction. This ensures that no unauthorized modifications can be done to data even by authorized personnel.

Encryption

Encryption is an important cybersecurity countermeasure that protects both data at rest and the data in transit. Encryption basically scrambles data using a "key" and makes the data unreadable for unauthorized users. During the encryption process, an authorized user utilizes a key with a specific encryption algorithm to make the "plaintext" data unreadable. On the other end, during the decryption process, the same user or another authorized user employs the same key with the same encryption algorithm to decrypt the "ciphertext". This means any unauthorized user cannot read the data transmitted until he or she has the key to decrypt the ciphertext.

An example of such encryption methodology is the Secure Sockets Layer (SSL). This is the data encryption methodology used by many websites to protect vital user data by preventing attackers from accessing sensitive user data that is moving to and from the website. This includes such critical data as user credentials, banking transactions and tax information. As SSL has become "breakable", it has now been replaced by Transport Layer Security (TLS).

Logging

System logging is a critical task of corporate technology environment. This is essential for traceability of user actions and events on any system. Most Cyber standards & regulations such as ISO 21434, PCI-DSS, SOX, ISO 27001 and others require the enabling of system logging as one of the controls required for compliance to that framework. Any organization using technology must constantly be logging and should do so by investing in a reliable logging solution. The organization must define what needs to be collected and the parameters needed to analyze this data. Also, an archive policy must be defined that enumerates what to archive and how long to archive that data. Further traceability is ensured by providing time stamps and certificates. Finally, in case the organization needs to follow specific Cyber standards or regulations, the accessibility to sensitive data also needs to be regulated.

Security Audits

Security audit is an important component for securing a technology system. A security audit is carried out to evaluate the functioning of a system and to determine the conformity or nonconformity of the system elements with any specified cybersecurity requirements or standards. Security Audits are needed to determine whether the implemented system is effectively meeting the specified cybersecurity objectives. This offers an opportunity for improvement to the system's cybersecurity profile and ensures that statutory & regulatory requirements are met for operating that system.

Fail Secure

Fail Secure means that if a device fails for any reason, it must fail in a manner which secures any data or environment it protects. A good example of this principle is the application of Fail Secure to a perimeter protection device such as a firewall. If a firewall fails in a manner (open) that is not secure, information external to the boundary protection device may enter, or the device may permit unauthorized information to be released to external parties. Thus, Fail Secure must ensure that the firewall fails in a manner that all traffic will be subsequently denied.

Defense in Depth

Any system should have a requirement to ensure whether the appropriate security measures are in place to prevent systems and networks from being compromised. Unfortunately, there is no single method that can successfully protect against every single type of attack. This is where the Defense in Depth architecture comes into play. Defense in Depth is a multi-layered defensive mechanism implemented to protect valuable data and information. If one mechanism fails, there is another layered mechanism that can also thwart an attack. This approach with intentional redundancies increases the cybersecurity of a system as a whole and addresses many different attack vectors.

Security By Design

Security by Design is an important concept in Cybersecurity. It means that organizations start building Cybersecurity into their product in the design stage of the development effort. Indeed, this should be done even earlier, at the concept stage, to ensure that Cyber best practices and solutions are employed to ensure that the system / component is secure.

Security by Design is focused on a proactive approach by preventing vulnerabilities that might be used by attackers to gain access into the system. This is opposite to a reactive approach in which organizations only try to respond and recover from Cyber attacks without any previous preparation or secure software development. However, in order to ensure Security By Design, an organization must have a culture of Cybersecurity ingrained into its very fabric. Only then, the organization shall ensure that Cybersecurity best practices are followed at every level of the organization and all stakeholders involved in enabling the Cybersecurity profile of the organization are onboard.

Privacy: A major concern in the Automotive sector

Privacy deserves a special mention when it comes to general cybersecurity terms as it is often differentiated from security of data. Privacy basically means the level of confidentiality a specific user's personal information is provided. Privacy of personal sensitive and private data, communications, & preferences is protected through several techniques & technologies and is influenced by many varying factors such as jurisdiction, privacy laws, type of data asset and security controls implemented. There are many regulatory & ethical Privacy frameworks that define requirements to protect individuals from substantial harm, embarrassment, or inconvenience due to the inappropriate collection, storage, or dissemination of personal information.

The Automotive sector has many major privacy concerns as great amounts of Personally Identifiable Information (PII) is collected by various stakeholders in the automotive sector. These include names, addresses, dates of birth, financial information, and user preferences. Apps in Connected Vehicles are also tracking user behavior and geolocation data for insurance, navigational guidance, and driver incentive purposes. This means that privacy has become an even bigger concern as Mobility undergoes the CASE transformation, as defined in Chapter 2 of this book.

Hacking & its Impact

Although, hacking is just one of several risks that Cyber Governance is used to manage, there can be no conversation on Cybersecurity without the mention of an organization being hacked. The organizations usually mentioned are highly visible brands and leaders in their own industry sectors. However, every day, thousands of smaller entities & individuals are hacked, and critical data assets stolen. Therefore, it is imperative for a Cyber SME to understand the various forms of hacking and its impact on an automotive organization.

Merriam-Webster dictionary defines Hacking[2] as "to gain illegal access to (a computer network, system, etc.)". However, a more precise definition of Hacking is the "attempt by an unauthorized individual to gain access to a computer system

---

[2] https://www.merriam-webster.com/dictionary/hack

or network by exploiting the vulnerabilities present in that computer system or network". Hackers are the individuals who attempt and succeed in entering a computer system or network through the act of hacking.

Usually, hackers are differentiated between Blackhat & Whitehat hackers. Blackhat hackers use hacking to commit cyber crimes for personal financial gain by means of cyber espionage or cyber extortion. Whitehat hackers, on the other hand, provide "Ethical Hacking" services to proactively protect an organization's critical computing assets. Based on the above differentiation of hackers, a more accurate characterization of hacking would be "extensive and detailed technical knowledge about computers and networks that enable an individual to bypass the system security built into the computing environment".

The below section explores many different forms of hacking techniques that are used to impact the A-I-C Triad of any computing asset.

Malware

Malware is short for "malicious software," and describes any malicious software, program or code that can cause harm to a computing resource. Malware is designed to be intrusive and is hostile in its intentions as it has severe negative impact on the performance of a computer system. Malware can be used to damage or take over control of computers, networks, tablets, and mobile devices.

Virus

Like a human or animal virus, a computer virus is a form of malware that can self-replicate itself by copying itself to another program. Hackers use computer viruses to infect vulnerable systems to gain control of the target system and steal user data. Viruses are usually propagated by tricking a trusting user to take an action that copies the virus into that individual's computer. This could be done by the user visiting an infected website or by opening an attachment in an email. It could also be enabled by clicking on an executable file or by viewing an infected advertisement on the Internet. Another way for a computer virus to propagate itself is by connecting to a USB device that is already infected by a virus.

Ransomware

The US governments' Cybersecurity and Infrastructure Security Agency (CISA) describes Ransomware[3] as "the type of malicious software, or malware, designed to deny access to a computer system or data until a ransom is paid". Ransomware is easy to protect against by having a Cyber mindset in the organization through a comprehensive & continuous User Cyber Awareness Program. Ransomware is only successful with the cyber-extortion of users & organizations who do not take cybersecurity seriously and do not follow cyber best practices.

Phishing

Phishing is a cyber attack that uses email to trick the email recipient into accepting the authenticity of the message in the email body. This trust makes the user click a link or download an attachment to the email. Phishing is, in fact, a form of spoofing i.e., a disguised form of communication that shows the communication from an unknown source to come from a known, trusted source. An example of a phishing attack is an email that is spoofed to be a request from a user's bank asking the user to click on a specific link to update contact information.

There are many forms of phishing attacks. Spear phishing is a form of phishing that is used to send tailored messages to targeted individuals or organizations. This is more sophisticated than sending out mass phishing attacks that might ensnare any user on the internet. An even more specific form of phishing attack is a whaling attack. The whaling phishing attack targets high-profile employees of a target organization, such as senior company executives, to steal sensitive confidential information from that specific company.

Finally, there is the Smishing attack. Smishing is short for "SMS phishing". Even though, it uses SMS messages as its attack vector rather than emails, it is still a form of a phishing attack. The aim in the case of a Smishing attack is for the user to be tricked into downloading malware onto his or her mobile device.

---

[3] https://www.us-cert.gov/Ransomware

Denial of Service

Denial of Service attacks are a specific type of hacking attacks that aim to overwhelm the resources of the target system so that authorized users can no longer use the target system for normal use. An example of such an attack is the Smurf DoS attack that utilizes ICMP traffic to overwhelm & block the network bandwidth of a targeted organization and thus make organizational resources unreachable.

In a Smurf Attack, the initial ping request to the destination has a spoofed source IP address of the target organization. This means that all ping responses from any system would be directed to the apparent source of the packet, i.e., the spoofed address of the target. As the number of machines replying to the ping request increase, the number of response packets attacking the target increase.

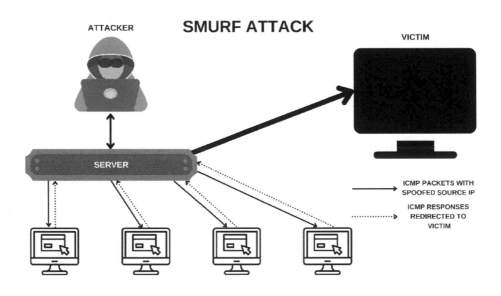

Figure 7 Smurf Attack

The intensity of a Denial of Service (DDOS) attack can be increased manifold by using a Distributed DoS (DDoS) attack. In this method, Botnets are used to intensify the attack on the target system or network. Botnets are basically a collection of compromised Internet-connected devices whose control has been taken over by the hacker. These could be computers, smartphones or IoT devices (vehicles) which had vulnerabilities due to non-compliance with security best practices. A nightmare scenario would be using a million vehicles as bots to propagate a DDoS attack.

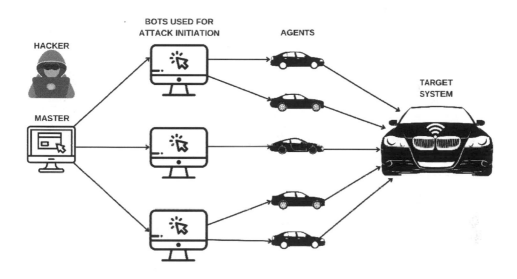

Figure 8 DDoS in Connected Vehicles

Vehicles are basically bots-on-wheels. A DDoS attack scenario can be war-gamed where millions of vehicles can be used to carry out a DDoS attack and degradation of the entire CAV eco-system made possible.

DNS Hijacking Attack

DNS Hijacking attacks are carried out by attackers when they change the DNS record of a target domain. The aim of DNS servers is to translate IP addresses and the networking info into human readable domain names. This enables an individual going to reputable sites such as yourbank.com to visit that site instead of trying to remember the IP addresses of these and millions of other websites on the Internet today.

There are several reasons why an attacker might carry out a DNS Hijacking attack. It could be used as another method of initiating a DoS attack. A DoS attack done via a DNS Hijacking attack enables attackers to compromise the target website and redirect legitimate users to an alternate system. Thus, authorized users cannot continue to use the resources of the original target system as intended.

A second reason why an attacker might want to use a DNS hijacking attack is to gather sensitive information. When legitimate traffic for a specific domain is redirected to a server of the attacker's choosing, this traffic redirection can be used to harvest user information that contain sensitive information. Such a system can be used as part of a phishing attack if the attacker can compromise the DNS information of the target organization.

Another illustration of collecting information via a DNS hijacking is the use of the target domain name for email communication. This not only includes capture of all in-bound email, but it can also be used by the attacker to send email masquerading as the target organization. This enables the attacker to cash in on the positive reputation of the target organization. An example of this is the use of the domain name of a reputable charity to defraud victims by collecting funds on behalf of the charity.

Insider threats

Insider threats should always be considered as a major threat by an organization while developing its Cyber Governance program. This is because insiders always have a higher level of access to data & resources and need to be trusted to perform their work in accordance with organizational policies and procedures.

However, a malicious actor can cause considerable damage if that individual is an insider. The impact could include Denial of Service of a service through physical manipulation or damage to the hardware or physical infrastructure of the organization. Furthermore, an organization may even suffer data breaches or information leakage through malicious internal actors through damage, theft, or loss of critical assets.

Cyber Governance, Risk & Compliance (GRC)

The impact of hacking and other Cyber incidents can be mitigated by implementing a comprehensive Cyber Governance, Risk & Compliance (GRC) program. The aim of such a program is to ensure that the organization takes a proactive approach towards enhancing its Cyber profile and ensures that a holistic Cyber Governance strategy is in place to enable the protection of data and other critical assets of the organization.

A comprehensive Cyber Governance program begins with planning of the organization structure and the roles of individuals who would be stakeholders in the effort to identify and secure the organization's data assets. It goes on to ensure that effective employment agreements are developed by the Human Resources department of the organization based on best practices for employee hiring. These best practices include background checks and job descriptions, security clearances, separation of duties & responsibilities, job rotation and termination practices.

It is critical to develop a high-level Cybersecurity policy for the entire organization that defines the vision of the organization with respect to the Cybersecurity profile of the organization. Furthermore, the organization needs to develop and use specific policies stating corporate position on various topics and the use of guidelines, standards, baselines & procedures to support those policies. Here, it is also important for the company employees to understand the differences between policies, guidelines, standards, baselines & procedures, and the priority of these documents in terms of their application to information security management in the organization.

From a human resource perspective, an important component is the Security Awareness Education for all employees of the organization. This is needed to make employees aware of the importance of cybersecurity, its significance, and the specific security related requirements relative to their position.

Data classification is also an important element of the Organization's Cyber Governance Program as data protection policies & controls can only be applied by Data Custodians based on the data classification assigned to a data element by its owner. This includes data elements that are sensitive, confidential, proprietary, private, critical, or even public information.

Risk Management

An essential part of a comprehensive Cyber Governance program is Risk Management, and it is crucial to develop a thorough set of risk management practices and tools to identify, rate, and reduce the risk to information assets of the organization. These best practices and tools should include the below areas of Risk Management:

- Asset identification and evaluation
- Threat identification and assessment
- Vulnerability and exposures identification and assessment
- Calculation of Single Occurrence loss and annual loss expectancy
- Safeguards and countermeasures

This topic will be explored in detail in Chapter 7 of this book.

Other Considerations

There are many other aspects to consider while developing a holistic Cyber Governance Program. These include the principles & controls that protect data against compromise and inadvertent disclosure and ensure the logical correctness of an information system; the consistency of data structures; and the accuracy, precision and completeness of data stored. Furthermore, the availability tenet of the A-I-C Triad must also be considered and the best practices that ensure that computer resources will be available to authorized users when they need it.

considerations include:

- The purpose of, and the process used for reviewing systems records, event logs and activities.
- The importance of managing change and its impact through a documented and controlled Change Management process.
- The application of commonly accepted best practices for system security administration, including the concepts of least privilege, separation of duties, job rotation, monitoring and incident response.

Finally, it is important to define the internal control standards that are required to satisfy obligations with respect to the law, safeguard the organization's assets and account for the accurate revenue and expense tracking.

What you Learned in this Chapter

The chapter has detailed the below:

- Understanding the A-I-C Triad and its criticality to the security of any data asset
- Enumeration of Privacy as it has great impact on Automotive Cybersecurity
- An overview of the key cybersecurity concepts such as providing Least privileges for the employees, Separation of duties, Encryption and Defense in Depth.
- The different Cyber threats including malwares and attacks like DOS, DNS Hijacking attacks that compromises the Organization's assets.
- The importance of the Cyber Governance, Risk & Compliance implementation in an organization to safeguard the valuable assets and data of the company.

Chapter 4: Cybersecurity Challenges in CAVs

Chapter Overview

This chapter explores the cybersecurity challenges that have manifested themselves by the development of Connected & Autonomous Vehicles (CAVs) and the conditions for enabling a secure and protected environment for CAVs. Many of these cybersecurity challenges are unknown as CAVs are still in their early developmental and experimental phase. However, the current cybersecurity gaps must be identified and need to be mitigated before CAVs can be safely and securely deployed in large numbers on our roads. Cyber SMEs can start by studying the In-Vehicle architecture and securing it. In addition, cyber domains related to effective operations of the Vehicle including cybersecurity monitoring of CAVs and incidence response protocols for breaches must also be investigated. In addition, a holistic approach should be outlined that evaluates the overall needs for Governance, Risk & Compliance for a CAV eco-system. Finally, the Cyber SME must also explore V2X security and determine the cybersecurity requirements of this new area of Automotive Cybersecurity.

Automotive Cybersecurity Defined

The transformation in the automotive sector towards enabling more technology in vehicles have been underway for almost a decade. This has meant more connectivity & automation for vehicles leading to the development of Connected & Autonomous Vehicles (CAVs). These vehicles pose unique cybersecurity challenges for all stakeholders of the vehicle. However, the current definition of cybersecurity does not fully address all aspects of automotive cybersecurity as it is only narrowly focused on Information Technology (IT) employed in any enterprise such as a financial institution, a Telco environment, or the head office of a manufacturing company. A more holistic definition of "Automotive Cybersecurity must be specified that encompasses all aspects of "people, processes &

technology" involved in the CAV eco-system. Having a well-articulated definition of Cybersecurity shall ensure that a better understanding of the Cyber risks to CAVs are enumerated. These are the Cyber risks that are exhibited by the adoption of this new technology and identifying the current cyber gaps in the CAV technology ecosystem.

Based on the above, Automotive Cybersecurity can be defined to encompass all the below:

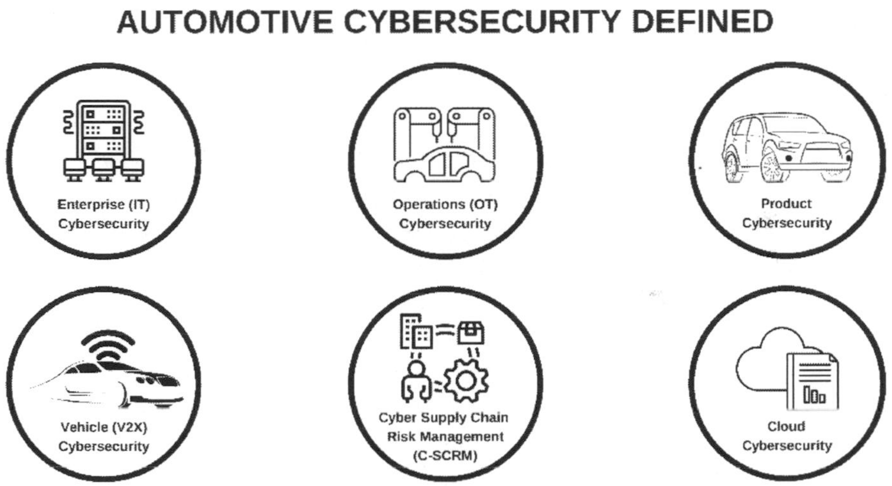

AUTOMOTIVE CYBERSECURITY DEFINED

Enterprise (IT) Cybersecurity

Operations (OT) Cybersecurity

Product Cybersecurity

Vehicle (V2X) Cybersecurity

Cyber Supply Chain Risk Management (C-SCRM)

Cloud Cybersecurity

Figure 9 Automotive Cybersecurity Defined

IT (Information Technology) Security: This is the traditional understanding of "Cybersecurity", also sometimes known as Information Security, and includes the controls within an enterprise that protect data and the flow of data for that organization. IT security is not only applicable in a manufacturing environment but also needs to be implemented in non-manufacturing environments such as Financial Institutions, Telcos, Government Organizations, Retail, and any other entity running information technology (IT) systems.

OT (Operation Technology) Security: OT Security is critical for any manufacturing environment. This discipline of Cybersecurity is focused on enabling security of Industrial Control Systems (ICS) such as SCADA, DCS and PLCs. Besides

53

manufacturing, OT security is also applicable in Petroleum Refining, Oil & Gas transportation, Mine Operations and Electricity generation, transmission & distribution.

In-Vehicle Security: This is an entirely new and innovative area of cybersecurity that needs to be developed to make certain that all the connected components in the vehicle operate in a secure manner.   This area includes Secure In-Vehicle Architecture, Encryption and Authentication.   It must also be ensured that the vehicle is being continuously monitored to detect any cyber events to respond to any malicious incidents.

Vehicle-to-Everything (V2X) Security: The Vehicle of the Future will be communicating with other vehicles (V2V), pedestrians (V2P), devices (V2D), infrastructure (V2I), cloud (V2C) and the grid (V2G). Data is being shared between all these entities and this communication needs to be secured to protect Personally Identifiable Information (PII). V2X Security will play an important role in privacy of commuters as less stringent security mechanisms would lead to location tracking and data compromise.

Supply Chain Security: For the Connected Vehicle, Supply Chain Cybersecurity will play a critical role in not only ensuring that a secure vehicle has been manufactured but shall also ensure that its operations are effectively managed. Integrations with multiple ECUs, patch updates and incident management, all require effective cooperation and collaboration between an OEM and its various suppliers.   Thus, having an effective Supplier Cyber Governance Program is essential for any OEM.

Cloud Security: In the days of the "Cloud", an automotive manufacturer or operator is continuously sending data to third-party service providers in the Cloud. Many of these are SaaS (Software as a Service) Apps that need to be monitored for the kind of data being uploaded to these third-party apps that may be hosting data in unknown jurisdictions. Cybersecurity tools are needed to protect this data and should be deployed by entities that send data from the Vehicle to the Cloud to ensure data security.

Thus, as defined above, Cyber Governance in the automotive industry is a vast and challenging undertaking. A holistic Automotive Cyber Strategy must evaluate and incorporate all the above-mentioned areas to ensure risk mitigation for all cyber threats to the CAV.

Cyber Risks to CAVs

A holistic approach to Cybersecurity in CAVs needs to begin with understanding the risks that connectivity and automation pose to consumers, developers & operators of CAVs. This major transformation in the traditional vehicle has occurred at a very rapid pace in a truly short duration of time. This has resulted in auto-tech solutions being implemented without recognizing the impact of cyber risks to CAVs that are prevalent in a connected environment. Only by comprehending these risks and their impact can we built a comprehensive approach to secure CAVs.

Some of the risks that need to be understood and safeguarded against are detailed below.

Protection of PII

Insight into the impact of risk starts by defining the assets that are in-scope and recognizing the criticality of these in-scope assets. The most significant asset in today's connected world is "data" and the collection, storage, and transmission of data in any connected environment poses significant cyber risk. CAVs collect immense amounts of data and a significant portion of this data is Personally Identifiable Information (PII) that is regulated under various frameworks and regulations. In addition, this data is being collected by various in-vehicle and 3rd Party apps. This implies that different vendors are managing this data by separate methodologies and there is no standardized approach implemented across the board to manage the security of this data.

Impact of Cyber attacks

One of the major drivers of implementing cyber solutions in CAVs is the ongoing & continuous risk of hacking to the vehicle itself. Such an attack can be disruptive to safe driving by disabling the engine while the vehicle is on the road or tricking an app software into changing lanes while another vehicle is in that lane. Thus, Cyber is not only a security issue but also a Heath & Safety issue for CAVs. This is different from Cyber in Financial Institutions where a breach might impact the financial well-being of its customers but will not lead to physical body harm. This means that public safety of all users on the road is a key concern of regulators for the CAV eco-system.

There are many possible Cyber attacks on the CAV eco-system. One nightmare scenario is that of 10,000 or 100,000 or even millions of CAVs being hacked together across different jurisdictions. These cyber attacks have been war-gamed as use cases and these are not just sci-fi scenarios. White-hat hackers have already publicly demonstrated many high-risk vulnerabilities in CAVs. One of the most famous of these was the example of a Jeep Cherokee[1] being taken over remotely through a vulnerability in its Uconnect system which resulted in a recall of 1.4 million vehicles. This means that Cyber risk has both financial and reputational impact on the OEM, besides being a Health & Safety issue for the drivers of the vehicles.

Security of the entire CAV In-Vehicle Eco-system

It is important to consider the segmentation of sub-systems to ensure the security of the entire In-Vehicle Eco-system. A modern vehicle consists of a large array of inter-connected systems that have security dependency on the entire eco-system. Many CAVs today have over a 100 ECUs (Electronic Control Units) and more than 100 million lines of code. This means that a vulnerability in one sub-system can be used to penetrate the entire In-vehicle eco-system. Also, many sub-systems are "connected" directly to outside entities which can lead to compromise of the system itself.

An example of such a vulnerable system is the infotainment app which can use Bluetooth or Wi-Fi connectivity to mobile phones to access music stored on that phone. The compromise of such a non-critical system itself is not a serious threat to the safety of the vehicle or its passengers. However, it can be used as a gateway into the entire In-Vehicle eco-system and a potential attacker might use such an exploit to move from non-critical to critical vehicle sub-system. Attackers have already successfully demonstrated that

---

[1] https://www.wired.com/2015/07/hackers-remotely-kill-jeep-highway/

inadequate segmentation between component systems can be used to move from low-risk apps to high-risk sub-systems that had serious impact on the functioning of the vehicle.

Patch updates as an Attack Vector

Over-the-Air (OTA) patch updates have become a critical functionality to keep software in sophisticated CAVs up-to-date. This is done to quickly update new vulnerabilities in software implemented in the vehicle as taking millions of vehicles to a dealer just for a software update is a major logistical challenge. However, recent cyberattacks have shown that OTA patching updates themselves can be used as an attack vector by using a Man-in-the-Middle attack. In an automotive industry with such a complex and diverse supply chain, it is imperative that ownership of software updates for any ECU or apps used in a specific vehicle is clearly demarcated by the OEM and a secure process defined.

Integrity of CAV Supply Chain

A modern vehicle is a complex machine with over 20,000 suppliers. This means that identifying all the risks associated with the components from these suppliers, the integration of these components and the entire supplier chain is a major undertaking. The OEM is ultimately responsible for the security of the vehicle but must deal with the current fragmented supply chain, lack of industry standards and well-documented poor security of ECU devices. This means that the OEM must adopt a multi-layered and comprehensive approach to cyber governance for the roll-out and operations of CAVs.

Privacy & Tracking

The impact to Privacy & Tracking of vehicle owners and operators is a major risk to be evaluated while doing a risk assessment. Telematics and location data are being used by many automotive-related businesses to improve their offerings. Examples of such businesses range from insurers who offer discounts based on safe driving to fleet owners who monitor usage to provide incentives to drivers of their fleet vehicles. However, that same data can also be used to identify individuals and monitor their movement, which raises numerous privacy concerns that need to be adequately addressed.

Current In-Vehicle Security Architecture in CAVs

The current In-Vehicle Security Architecture (IVSA) in CAVs have significantly evolved since the early days and today, there is significant effort in building technology that secures the flow of data in the vehicle. The building blocks of the In-Vehicle Security Architecture remains Electronic Control Units (ECUs) and the CANBus. However, today many companies provide solutions around "Secure OS Wrappers", secure OTA Updates, encryption technologies and segmentation inside the vehicle technology eco-system.

A few basic details of the IVSA must be followed to ensure a secure architecture.

- Details of all the ECUs in the vehicle. An ECU or the electronic control unit is used in Connected Vehicles to control the functioning of vehicle components through a computer with internal pre-programmed and programmable computer chips. In a modern vehicle, there are more than a hundred ECUs including Airbag ECU, Steering & Braking ECU, Engine & Transmission ECU, ADAS ECU and many others.

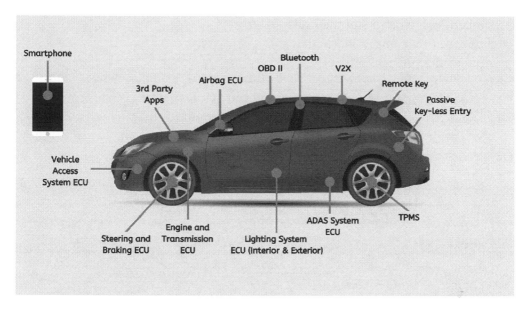

Figure 10 ECUs in a Vehicle

- The CANBus is short for Controller Area Network Bus. This is connectivity protocol that was developed by Robert Bosch and has quickly gained acceptance in the automotive and aerospace industries. It is the CANBus that connects the various ECUs together.
- All external connectivity points must be identified, and data transmitted via these connectivity points controlled and monitored.
- Segmentation requirements between various critical and non-critical apps & systems must be evaluated, implemented, and enforced.
- Encryption needs & methodology must be recognized and standardized.
- OTA updates for all software used in the vehicle must be securely managed through a standardized methodology. Ownership & reporting of this activity must be clearly defined.
- Authentication & Access Control requirements must be assessed and built into the vehicle for all its systems.
- Vehicle apps that communicate to servers in the Cloud must be cataloged and security technologies applied to monitor & control this transmission.
- Defense-in-Depth concept must be implemented to ensure a secure IVSA design.

Cybersecurity Monitoring for CAVs

CAVs are compute devices that process, store, and transmit immense volumes of data on an ongoing basis. These compute devices should be regularly monitored for Cybersecurity issues that enable its users to protect themselves from fraud, misuse, or breaches. This monitoring can be done through Vehicle Security Operations Centers (VSOCs).

In the Information Security world, it is a commonly implemented best practice for organizations to either operate a Security Operations Center (SOC) or to outsource this activity to a Managed Security Service Provider (MSSP). This is done to proactively detect any security incident on the computers, servers & networks it owns and operates. This proactive behavior ensures that the organization can prevent and respond effectively to cybersecurity threats.

Similarly, there is a strong need for Automotive owners to ensure that the vehicles they operate can proactively detect any security incident by effective and continuous cyber monitoring. This can be done through a VSOC. A VSOC has the following core functions:

- Ingest various relevant data feeds about CAV vulnerabilities
- Correlate between the various feeds
- Enable Mobility-specific analytics
- Enable Real-time detection of security incidents in CAVs
- Ensure Vehicle Cyber Incident Response Playbooks (VCIRP) are defined that elaborate specific Cybersecurity triggers and follow-up protocols
- Carry on 24x7 Cyber monitoring of vehicles in specific fleets

One big challenge today is that it is not clear who would be responsible for operating these VSOCs and monitoring millions of Connected Vehicles on our roads. This is because business models in the mobility sector are continuously evolving, and OEMs are exploring different options to ensure the security of CAVs. These options include operations of the VSOC by the OEMs themselves, by fleet owners or even by third-party Managed Vehicle Security Service Providers (MVSSPs).

Vehiqilla Inc.[2] was started to ensure that this critical gap in automotive cybersecurity is met. More details on Vehiqilla and its VSOC services are given as a case study in Chapter 10 of this book.

Vehicle Cyber Incident Response Playbooks for Managing CAV Fleets

Most organizations have a documented Cyber Incident Response Plan that is focused on how to handle security breaches in the Information (IT) Security domain. This same approach must be mirrored to ensure effective and immediate incidence response is possible if a CAV undergoes a Cyber incident. Thus, having well-defined Vehicle Cyber Incident Response Playbooks (VCIRPs) to manage cyber incidents for CAVs will need to become an integral part of the operational plans for any organization that operates a fleet of vehicles.

While developing a VCIRP, the three main objectives of any Incident Response Plan must also be applied to a VCIRP. The first objective of any VCIRP should be to detect and identify an attack to any vehicle in the fleet. The second objective of the VCIRP should be to define specific actions that need to be undertaken to contain the damage form the attack. The third and final step of the VCIRP would be to eradicate the root cause of the incident and ensure that the same vulnerabilities do not continue to remain in the CAV eco-system.

However, the implementation of Vehicle Cyber Incident Response for any fleet owner should have much broader goals in changing the mindset of the organization. These include:

- Define the Vehicle Cyber Incident Response Team (VCIRT).
- Prepare the Vehicle Cyber Incident Response Team (VCIRT) to deal with threats to its fleet of Connected Vehicles.
- Establish Methodology to detect & isolate cyber incidents in its fleet & identify their severity and impact.
- Define actions to stop the attack & eradicate the underlying cause. This could be a new patch or change in user behavior.
- Ensure Connected Vehicles are recovered to operational status.

---

[2] https://vehiqilla.com/

- Establish procedures to conduct a post-mortem analysis to learn from the incident & prevent future attacks to the CAV Fleet.

As mentioned above, an important objective of the Vehicle Cyber Incident Response Playbook is to prepare the Vehicle Cyber Incident Response Team (VCIRT) to deal with any cyber incident to an organization's CAV fleet. This means that it is imperative to have the correct expertise on the VCIRT and ensure that the right people are on-hand as part of the VCIRT to deal with the Incidence. There are some key roles that must be filled in an organization's VCIRT that owns or operates a CAV fleet. Below is a comprehensive list of such roles:

- Vehicle Cyber Incidence Response Manager
- C-level executives (ownership of key business decisions)
- CAV Cybersecurity analyst
- Auto-tech engineer
- CAV Threat researcher
- Legal representative
- Corporate communications
- Human resources
- Risk management
- External CAV Cybersecurity forensic experts

Finally, use cases and scenarios for Vehicle Cyber Incidents (VCI) must be defined and war-gamed before an actual real-life cyber incident happens. This shall ensure that all stakeholders understand the VCIRP and can initiate response plans to specific triggers.

Some major cyber incident scenarios that have been defined by ENISA[3] in their detailed guide ENISA: Cyber Security and Resilience of smart cars. These are:

- Remote attacks that threaten passenger's safety
- Persistent vehicle alteration by the legitimate user or by using diagnostic equipment
- Theft
- Surveillance

---

[3] https://www.enisa.europa.eu/publications/cyber-security-and-resilience-of-smart-cars

The Vehicle Cyber Incident Response Playbook should define well-articulated protocols that must be activated for each of these scenarios. In addition, these protocols must have been practiced & tested multiple times before any such incident takes place. This will ensure an effective and timely response to any Vehicle Cyber Incident (VCI) and protect the vehicles in the fleet from an untoward cyber incident.

Holistic Cyber Framework for CAV Governance, Risk & Compliance

Based on the above discussion, it has become clear that organizations developing, owning, or operating CAV fleets need to adopt & implement Cyber Frameworks developed specifically to manage Governance, Risk & Compliance in CAV fleets. Such frameworks fall into two broad capabilities:

- Risk Management Frameworks
- Cyber Program Management Frameworks

Risk Management Frameworks

A Risk Management Framework is a system of standards, guidelines, and best practices to manage risks that arise in the Online & Connected World we live in today. Such a framework would cover the following areas:

- o Risk Assessment Methodology
- o Identification, measurement & quantification of risk
- o Evaluation & recommendation of relevant Safeguard / Controls
- o Enumeration of a Risk Mitigation Strategy
- o Prioritization of Cyber activities related to mitigating the risks identified

Cyber Program Management Framework

A Cyber Program Management framework is a system of standards, guidelines, and best practices that provide the structure & methodology an organization needs to protect its data assets. A Cyber Program Management Framework would incorporate the following:

63

- Evaluate state of the Cyber Program implemented in the organization.
- Build a Holistic Cyber Program that ensures cybersecurity best practices are implemented and followed in the organization.
- Establish Baseline Controls to ensure a minimal level of cybersecurity.
- Ensure Measurement of KPIs to continuously update and evaluate the state of the Cyber Program.
- Communicate the Cyber Program Objectives to all internal and external stakeholders

CAV Cyber Frameworks

For the cyber management of CAV fleets, currently many different frameworks are being developed which can be applied to the above-mentioned aspects of Cyber Governance, Risk & Compliance.   The major cyber frameworks that could be applicable for any automotive eco-system are

- ISO 21434 Road Vehicles Cybersecurity Engineering Standard
- NIST Cyber Security Framework
- ENISA Good Practices for security of Smart Cars
- UNECE WP.29 R155 & R156 Cybersecurity Regulations
- Transport Canada Vehicle Cybersecurity Guidance
- TISAX
- ISO 27000 Series
- PCI DSS

These are explained in detail in Chapter 8 of this book.

V2X Security

Vehicle-to-everything (V2X) Communication is another critical area of Mobility as the world moves towards an age in which Connected & Autonomous Vehicles (CAVs) are ubiquitous. V2X would ensure direct communications with Intelligent Transport Systems (ITS). ITS ensure safety and efficient traffic flow in SmartCities and need successful communications with the CAVs to enable this smooth flow of traffic. Cyber-threats and attacks are a major challenge[4] to V2X as these can directly impact ITS efficiency, a variety of functionality and integrity.

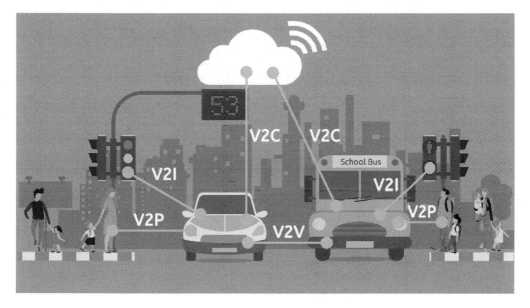

Figure 11 Vehicle to Everything (V2X)

---

[4] http://wrap.warwick.ac.uk/106474

Below is the terminology used with reference to V2X.

Table 2 Vehicle-to-Everything Terminology

Vehicle 2	Everything
V2I	Vehicle-to-Infrastructure
V2N	Vehicle-to-Network
V2P	Vehicle-to-Pedestrian
V2V	Vehicle-to-Vehicle
V2D	Vehicle-to-Device
V2G	Vehicle-to-Grid

V2X Communication Technology

V2X communication is based on two different wireless communication protocols. These are DSRC and C-V2X.

- DSRC was the original V2X wireless communication protocol. It stands for dedicated short-range communications. It was published by IEEE in 2012 and is also called ITS-G5 in Europe. It is based upon underlying radio communication provided by 802.11p and is used for direct communications between moving vehicles. Thus, it is independent of both the cloud & cellular infrastructure and supports V2V and V2I.
- C-V2X stands for cellular V2X and leverages cellular technology to enable the link between the vehicle and the rest of the world, including other vehicles and the traffic control system. It was published by 3GPP in 2016 and has V2X specifications based on LTE as the underlying technology. As it is cellular based, it supports V2N as well as V2V & V2I.

V2X Use Cases

V2X communications can be used to ensure that traffic flow is smooth and connected vehicles communicate their intentions to other vehicles, infrastructure, and devices around them. Some examples of possible use cases for V2X are

- Forward collision warning

- Lane change warning/ blind spot warning

- Emergency electric brake light warning

- Intersection movement assist

- Emergency vehicle approaching

- Roadworks warning

- Platooning

V2X Cybersecurity challenges

V2X communications raise several cyber challenges that need to be identified, understood, and managed. A critical area of Cyber is Governance and a holistic Cyber Governance framework needs to be established to manage and mitigate the risks associated with V2X communication. This means that an Intelligent Transport System (ITS) architecture needs to be implemented that is secure, protected and resilient. This architecture should ensure secure & reliable communications that incorporates Two-way Authentication and Cryptography to provide Confidentiality, Integrity & Availability for all V2X Communications. In addition, Privacy and Data Protection are two key areas that need to be addressed by all jurisdictions that enable V2X.

Security Credentials Management System (SCMS)

The Security Credential Management System leverages Public Key Infrastructure (PKI) to ensure the security of V2X communication. As per the United States

Department of Transport (US DoT), an SCMS[5] provides the following security features to a V2X environment:

- Integrity: This enables users to trust that the message was not modified between sender and receiver

- Authenticity: This enables users to trust that the message originates from a trustworthy and legitimate source

- Privacy: This enables users to trust that the message appropriately protects their privacy

- Interoperability: This ensures that different vehicle makes, and models will be able to talk to each other and exchange trusted data without pre-existing agreements or altering vehicle designs[6]

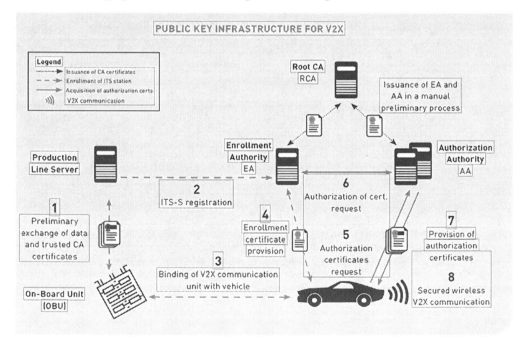

Figure 12 Public Key Infrastructure for V2X[7]

---

[5] https://www.its.dot.gov/resources/scms.htm
[6] https://www.thalesgroup.com/en/markets/digital-identity-and-security/iot/industries/automotive/use-cases/v2x
[7] https://iot-automotive.news/gemalto-v2x/

What you Learned in this Chapter
The chapter has discussed the following:

- Specifying the requirement for Cyber Governance in CAVs.
- Articulating a clear definition of Cybersecurity in CAVs.
- The major Cyber Risks to CAVs including exposure of PIIs, Cyber-attack vectors for CAVs, a vulnerability on a vehicle subsystem compromising the entire vehicle and the security issues caused through OTA.
- Details of the Current In-Vehicle Secure architecture.
- Core functions of VSOC to enable Cyber monitoring in CAVs to detect & prevent Cyber threats.
- Importance of holistic Cyber Framework implementation in the CAV fleets
- V2X Security, communication technology used and related Cyber challenges.

Chapter 5: OT Cybersecurity: Securing Manufacturing Networks

Chapter Overview

The objective of this chapter is to cover the architecture, processes and technologies involved in Operational Technology (OT). OT plays a critical role in the vehicle manufacturing eco-system and needs to be effectively secured. Beginning with the definition of various OT systems, this chapter goes on to define the architecture as well as the components of OT. It then delves into various aspects of OT Cybersecurity and further enumerates the cyber challenges that have manifested themselves because of the convergence of IT & OT.

What is Operational Technology

Operational Technology broadly defines the technology used in enabling vehicle manufacturing but also manufacturing in general. The more defined term given to OT is Industrial Control Systems (ICS). Besides manufacturing, ICS tools can be used to manage physical processes in many different industries, including Petroleum Refining, Oil & Gas transportation, Mine Operations and Electricity generation, transmission & distribution.

Generally, ICS Systems are divided into three main categories:

- Supervisory Control and Data Acquisition (SCADA)

  SCADA systems collect data from various collection devices that monitor and / or control a specific industrial task and then consolidate this information at a central server for historical records or visualization needs.

- Distributed Control Systems (DCS)

  DCS is a system that is used to control continuous or batch-oriented processes consisting of functionally and / or geographically distributed controllers and Input/Output (I/O) via networks for monitoring & communication.

- Programmable Logic Controllers (PLC)

PLCs are microprocessor-based devices used to control industrial processes or devices. They provide advanced functions including analog monitoring, control, and high-speed motion control as well as share data over communication networks.

ICS Architecture

The traditional ICS architecture is based on the communication between two main components i.e., Servers and Controllers.

Servers

- Control Server

The main task of the control server is to connect with specific control devices across an ICS network. This is done using supervisory control software that establishes and maintains this communication.

- Input / Output Server

The I/O server is responsible for the collection of process information from sub-components (PLCs, RTUs and IEDs). This means that I/O servers can be used to interact with third-party control components such as a control server or HMI.

- Master Terminal Unit

In an ICS network, PLC devices and remote telemetry units (RTUs) are usually dispersed. The MTU is used to send commands to RTUs in the field and operates as a master server in such an ICS network. An MTU is sometimes also referred to as a SCADA.

- Data Historians

Data Historians are systems that log all processed data within an ICS and thus, can also help in performing different tasks including statistical process control, process analysis, planning and report functions. In such a centralized database, each measured item is assigned a tagname, timestamp, value and data quality indicator. Historians log data records for all these data points and store them in the form of a series of secure binary files that provide the option of fast retrieval. Such Log data is usually sent to the corporate IT solutions to provide correlation with other organizational activities.

Controllers

- Remote Telemetry Unit

The objective of an RTU is to support data acquisition and control in remote SCADA stations. Thus, these are field devices outfitted with wireless and wired interfaces that enable an RTU to receive commands and send data back to the MTU.

- Programmable Logic Controller

A PLC works as a local controller in DCS and performs many activities in the system. PLCs can enable a wide range of input and output signals to sense environments and monitor and control industrial machines. Basically, a PLC provides management guidance from control devices such as actuators and sensors. It keeps data and operating procedures in their memory and will transfer data only if triggered by the MTU.

- Intelligent Electronic Devices

IEDs are a critical part of ICS as they enable control at the local level in an automated manner. This is because these are intelligent or smart enough to perform logical processing and control. They can collect various kinds of data from the ambient environment, communicate it to other devices and conduct local processing and control.

- Programmable Automation Controller

A PAC basically combines the capabilities of a computer with the functionality of a PLC, thus providing higher-level instructions. Although most PLCs in today's world are remarkably similar to PACs, PACs are more advanced from PLCs due to their more open architecture and modular design. A major difference between PACs and PLCs is that PACs can control & monitor a larger number of complex or high-speed analog I/O. Thus, in an environment where advanced automation is required, such as a complex automation system or a large processing plant, PACs are a better choice than PLCs.

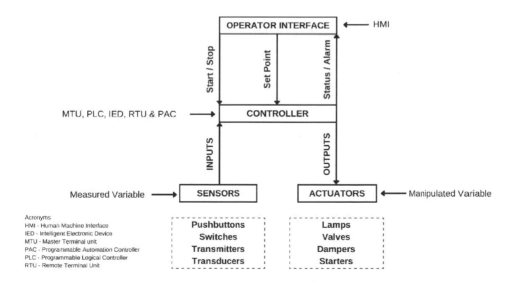

Figure 13 OT Components

Other ICS Components

- Human Machine Interface

An HMI enables interaction between humans and machines through software and hardware components. HMIs provide the ability for operators in control rooms to monitor the entire chain of processes under control, make configurations changes, alter control variables, and exercise local processing and control. HMIs also show status information and historical data amassed by all devices in the ICS system in an easy-to-understand graphical user interface (GUI). Finally, another critical function of HMIs is to ensure that Operators can use the HMI to issue manual overrides in case of an emergency.

- Sensors

Sensors are basically an input device that enables an output (signal) with respect to a specific physical quantity (input). It can also be said that a sensor is a device that converts signals from one energy domain to electrical domain.

There are many different types of sensors:

- Temperature Sensor
- Proximity Sensor
- Accelerometer
- IR Sensor (Infrared Sensor)
- Pressure Sensor
- Light Sensor
- Ultrasonic Sensor
- Smoke, Gas and Alcohol Sensor
- Touch Sensor
- Color Sensor
- Humidity Sensor

- o Tilt Sensor
- o Flow and Level Sensor

- Actuators

An actuator is a component in any machine that enables movement i.e. it helps a machine achieve physical movements by converting energy. Actuators are mechanical or electro-mechanical devices that provide controlled movements stimulated by electrical or manual impulses. This stimulation can also happen through other means such as the flow of fluids such as air, hydraulic, etc.

There are two basic motions that actuators enable i.e., linear and rotary. Linear actuators convert energy into straight line motions and usually have a push and pull function. On the other hand, rotary actuators convert energy to provide rotary motion. A typical use is the control of various valves such as a ball valves or butterfly valves.

Below are a few categories of Actuators based on their stimulation impulse and type of motion:

- o Electric Linear
- o Electric Rotary
- o Fluid Power Linear
- o Fluid Power Rotary
- o Linear Chain Actuators
- o Manual Linear
- o Manual Rotary

OT Cybersecurity defined

Legacy OT systems have not been designed or implemented with security in mind. They are often proprietary and have not been tested for security vulnerabilities. They also do not integrate with newer versions of Operating Systems, leaving backdoors through which hackers can exploit the system. This means that although Legacy OT systems have significant strengths, they have many weaknesses from the Cybersecurity perspective. Legacy OT systems strengths include having a high degree of Availability and a high degree of Authorization while the two primary cybersecurity weaknesses are allowing Total Authorization and Total Trust.

There have been many recent instances of the impact of Cyber breaches on an ICS. The most prominent of these Cyber breaches was the 2015 Ukrainian hack[1] that shut down power to 230,000 customers by remotely opening breakers. This was done through the NotPetya ransomware and led to a global wave of breakdowns. Besides the Ukrainian energy provider, others impacted in the same manner included drug giant Merck, legal firm DLA Piper and global shipping business Maersk.

Internet-enabed HMIs are also another target of hackers. Connecting HMIs to the Internet allows remote connectivity and greatly increases the convenience to perform various tasks. However, it comes at a price. This was proven by the BlackEnergy 2 Malware[2] which was able to target internet-enabled HMIs that had such added ICS specific features. An important thing to keep in mind is that in many such breaches of industrial espionage, the goal might not be disruption but reconnaissance and gathering of more information of the ICS environment.

IT & OT Convergence

Traditionally, completely distinct, and separate organizational departments managed Information Technology (IT) and Operational Technology (OT) without any interdependence on one another. This has slowly changed during the recent times with various "IT" technologies being utilized to enhance the performance of the OT eco-system. This includes using TCP/IP & Ethernet for connectivity, usage of Windows & Unix Servers, Oracle & SQL databases and Webservers to provide access from the Internet. The advent of the Cloud and Work-from-Home (WFH) has further accelerated this paradigm shift.

---

[1] https://www.forbes.com/sites/thomasbrewster/2017/07/03/russia-suspect-in-ransomware-attacks-says-ukraine/#7a084ee96b89
[2] https://www.pcworld.com/article/2840612/attack-campaign-infects-industrial-control-systems-with-blackenergy-malware.html

The merger of IT with OT is driven by the need to optimize the collection and exchange of data between machines, infrastructure assets & applications while interoperable scaling processes across physical and virtual systems. This integration promises numerous benefits including improved flow of information, process automation, advances in the management of distributed operations and better adherence to regulatory compliance. However, this convergence of IT and OT has led to OT systems being even more vulnerable to Cyber attacks in today's connected world. The main reasons why the risks have increased for OT Systems are:

- Adoption of standardized technology with known vulnerabilities
- Connectivity of Control systems to other networks
- Constraints on the use of existing security technologies and practices
- Insecure remote connections amid widespread remote access availability
- Widespread availability of technical information about control systems

Risk-based OT Cyber Governance

The billion-dollar question is how the industry can ensure that such OT-based systems are protected from cyber breaches. The basics tenets of cybersecurity need to be applied here i.e., enabling the confidentiality, integrity, and availability of such systems, and ensuring that sensitive data is not compromised. The first step in this regard is to have a single owner of all Cyber Governance, Risk & Compliance initiatives in the organization. The Chief Information Security Officer (CISO) role is independent of the IT or the OT department. The individual must be a C-level functionary and must report directly to the Chief Executive Officer (CEO) independent of the Chief Information Officer. If the CISO reports to the CIO, this makes the CISO part of the IT department of the organization and this does not align with the overall business requirements of a manufacturing organization.

Besides ensuring there are various other Cybersecurity Policies in place, the CISO must also have a defined OT Cybersecurity Policy. This policy must define the Acceptable use of SCADA systems, the data flow permissions, the Access Control Mechanisms for console or remote connectivity, the audit trail requirements and the cyber monitoring needed to detect any harmful occurrence. The CISO must also build a team of ICS Cyber SMEs who are adequately trained and understand ICS Cyber challenges.

Securing Future Industrial Networks: Industry 4.0

Industry 4.0 or the Fourth Industrial Revolution (4IR) will revolutionize manufacturing processes by using the combination of automation, intelligence, and the Internet of Things (IoT). This will deliver greater speed, agility and use innovation to enhance all aspects of manufacturing and supply chain, thus greatly adding value to human life.

Industrial IoT (IIoT) is a good example of this change brought to manufacturing through Industry 4.0. IIoT is a matrix of interconnected sensors, instruments, and devices that collect & share data that can be leveraged across industries, such as manufacturing, oil and gas, transportation, energy/utilities. IIoT can do this through converged IT/OT ecosystems serving as conduits that deploy IIoT into the Industry 4.0 ecosystem. This data can then be used to analyse and predict metrics that can be applied to enhance performance. However, this does mean that there is greater Cyber risk for these inter-connected organizations and industries.

Business leaders can ensure that their organizations and, indeed their entire industry sector, has adequate cyber governance in place by ensuring:

- Make Cybersecurity the highest priority of the organization by having a Chief Information Security Officer (CISO) responsible for all Cyber Governance, Risk and Compliance and having the CISO report directly to the CEO.
- Use Cyber Innovation to quickly incorporate new Cyber processes & technologies to protect against emerging Zero-day Threats.
- Build a Zero Trust architecture that only allows authenticated users to use authorized applications on authorized devices.
- Define the critical assets of the organization that must be protected in the perimeter-less connected world of the Cloud & WFH.
- Make Cybersecurity the culture of the organization through a defined and continuous Cybersecurity Awareness Program.

- Ensure that Cybersecurity is considered during the implementation of any new technologies, processes, or procedures i.e., use Security-by-Design principles.

What you Learned in this Chapter

The chapter provides details on the following areas of OT:

- Basic definition of Operational Technology (OT) or Industrial Control Systems (ICS).
- An overview about different technologies and their categories involved in the Operational Technology (OT).
- Explanation of the ICS architecture describing the different ICS components including the various types of servers and controllers involved in the ICS network.
- The Cybersecurity Challenges in a legacy OT environment.
- The convergence of IT & OT.
- The Cybersecurity challenges manifesting themselves in the converged OT-IT eco-system.
- The holistic methodology that can be used to secure Industrial networks with adequate Cyber governance.

Section 3: Cyber Solutions for securing CAVs

Chapter 6: Building Cyber Governance Programs for Automotive Organizations

Chapter Overview

The objective of the chapter is to provide a deep overview about the importance of Cyber Governance for an Automotive Organization. The chapter begins with detailing the criticality of Corporate Governance programs in general and then goes on to give a holistic definition of Cyber Governance. It lays out the building blocks of a comprehensive Cyber Governance Program and finally highlights the KPIs that need to be measured to evaluate an effective Cyber Governance Program.

Corporate Governance

In today's competitive landscape, the secret to market success for any corporate organization is a well-formulated and disciplined corporate governance framework. Having a strong foundation of Corporate Governance is also true for the Automotive sector with its comprehensive supply chain. This eco-system includes defined relationships between Original Equipment Manufacturers (OEMs) to Tier 1 & Tier 2 Suppliers. It further includes relationships with other stakeholders such as dealers, distributor networks and post-sale maintenance & operations.

The aim of Corporate Governance is to protect the interests and requirements of a company's many stakeholders including shareholders & investors, employees, customers, partners, and the society itself. This protection is guaranteed through a well-defined and robust Corporate Governance framework that is comprised of standards, policies, processes, practices, and procedures. All Corporate Governance frameworks follow certain basic principles given below:

- Clearly Defined Strategic Objectives
- Organization Discipline through an establishment Governance Framework
- Effective Risk Management
- Protecting the interests of the Employee & the Customer
- Ownership, Accountability & Transparency
- Continuous Improvement

As companies in the automotive sector become more and more reliant on technology, the frequency and range of Cyber Threats to their businesses has greatly increased. This enhanced impact of cyber means that there is a critical need to incorporate Cyber as part of an organization's Corporate Governance Program. Such a Cyber Governance program should be holistic in nature and should ensure that all principles of good Corporate Governance are also supported in the Cyber Governance Program.

Cyber Governance Defined

The main aim of a Cyber Governance Program is to define the ownership of all aspects of an organization's Cyber initiatives. It also identifies the structure that would be used to implement and operate this Cyber Governance program in the organization. For the companies in the Automotive sector, the Implementation of a holistic Cyber Governance Program shall ensure that the organization's board is ultimately responsible for Cyber risk management. It also provides an effective line-of-sight for the adoption of standards, processes, and best practices that, together, secure the organization against cyber risk.

A holistic Cyber Governance Program must identify all areas in which the board should act to improve its cyber risk management. These areas include:

- Cybersecurity Strategy
- Cybersecurity Culture Change
- Cybersecurity Organizational Structure
- Cyber Asset Management
- Cyber Risk Management

- Legal, Regulatory and Contractual Obligations
- Cybersecurity Management Framework
- Cybersecurity Controls
- Business Continuity and Incident Management
- Employee Cyber Awareness Training Program
- Vulnerability Assessment & Patch Management
- Service Provider (Supplier) Risk Management Program

Building a Cyber Governance Program

The below section enumerates the major building blocks for enabling a holistic corporate Cyber Governance Program in the organization.

Cybersecurity Strategy

The Oxford dictionary [1] defines the term "Strategy" as "A plan of action designed to achieve a long-term or overall aim" while the Business Dictionary[2] defines "Strategy" as "a method or plan chosen to bring about a desired future, such as achievement of a goal or solution to a problem". Thus, every organization in the Automotive sector needs to start by initiating a discussion at the highest levels on the necessity of having a well-defined Cybersecurity Strategy for their organization.

The Cybersecurity Strategy is in fact an enumeration of the organizational vision and a mission statement of the organization that declares the objectives of the organization and the importance of Cybersecurity in achieving those objectives. Thus, the Cybersecurity Strategy must set the direction for the entire organization's Cybersecurity efforts. A few key questions must be answered during the formulation of the Cybersecurity Strategy:

- What is the Organizational Vision with regards to Cybersecurity?

---

[1] https://www.lexico.com/definition/strategy
[2] http://www.businessdictionary.com/definition/strategy.html#:~:text=SUBJECTS-,strategy,See%20also%20tactics

83

- What are the business drivers for enabling Cybersecurity in the Organization?
- What are the Organizational Strengths & Weaknesses when it comes to Cybersecurity?
- What are the Threats & Opportunities to the organization in the field of Cybersecurity?
- Can the Organization leverage external factors to excel in business by enabling Cybersecurity Best practices?
- What are the Critical Success Factors needed for the Organization to thrive in enabling a successful Cybersecurity Program?

The below image gives a holistic overview of the various aspects of a detailed Cybersecurity Strategy. It explores the purpose, the vision, the capability roadmap, the information security frameworks applicable, the core investments needed and the operating objectives for enabling such a holistic Cybersecurity Strategy in the organization.

Figure 14 Cybersecurity Vision & Strategy

The Organization can carry out a SWOT (Strengths-Weaknesses-Opportunities-Threats) Analysis to answer many of the questions pertaining to the over all strategy. This is a method that was designed by Albert Humphrey for the Stanford Research Institute in the 1960s and can easily be adapted to explore Cybersecurity Strengths, Weaknesses, Threats & Opportunities for an Automotive Organization.

Cybersecurity Culture Change

One extremely important component of Building a Cybersecurity Strategic plan is to ensure a change in the culture of the Organization with regards to Cybersecurity. Ultimately, Cyber is a human mindset challenge. Changing the culture for Cybersecurity for the entire organization through a successful Cybersecurity Strategy is not an easy task. This is because Cyber must be "baked" into the culture, attitude, and intention of ALL the stakeholders, including the Board of Directors, executive management, employees, shareholders, suppliers and, indeed, the customers. If this is not done, the Organization will not meet the mission objectives laid out in the Cybersecurity Strategy.

This change of the Cybersecurity culture must start at the very top. The Board of Directors (BoD) is ultimately responsible for all Cyber initiatives in the organization and thus, must set the agenda that every stakeholder must abide by. In addition, the Board of Directors also has the ownership of oversight of the Cyber hygiene of the organization. Thus, it is essential that Cyber remains on the Board's radar on a continuous basis.

In addition, all the employees need to understand that Cyber is "everyone's responsibility" throughout the organization. Corporate Cyber Policies & objectives must be clearly defined and adherence to these must be on every performance plan and annual appraisal. The aim of a meticulously designed and effective Cyber Governance Program is to produce a culture where everyone in the organization recognizes and executes their role in protecting the confidentiality, integrity and availability of the organization's crown jewels.

However, building a Cybersecurity Culture must go well beyond encouraging employees to follow Cyber best practices as per the Employee Cyber Awareness Training Program. The Cyber Innovation Leadership Framework[3] shows that a mature Cybersecurity Organization is still limited in its outlook. The organization can truly survive and thrive in the connected world only when everyone "owns"

---

[3] https://www.cyberinnovationleadership.com/cyber-innovation-leadership-framework.php

cybersecurity in the organization that enables creation of innovative solutions to protect the Organization from a Cybersecurity Incident.

Some principles for enabling a Cyber Culture Change in an organization are:

- Everyone OWNS Cybersecurity
- Cybersecurity Awareness must be more than just a training exercise
- Secure Application Development is critical to the Organization's Cybersecurity Profile
- The reward structure of the organization should incentivize Cybersecurity Best Practices
- The organization must create and engage a Cybersecurity Community internally.

Cybersecurity Organizational Structure

For any organization, the ownership of Cyber Governance must ultimately lie with a business function. This becomes even more important for an automotive organization as Cyber incidents can directly impact business and lead to considerable loss of reputation. This is because a Cyber breach that makes technology systems unavailable during manufacturing has many distinct consequences. First, it would drive down business revenue as no vehicles will roll out of the assembly line during the downtime through loss of production. Second, and more seriously, it would also affect the reputation of the company as a Cyber incident might lead to recall of vehicles. This might make existing & potential customers turn to other competitors as they no longer trust the organization with their confidential data. Finally, this loss of reputation can have a direct effect on the share prices of the Automotive Organization due to loss in investor confidence. Thus, defining ownership of the Cyber Governance program is essential and a critical first step to the success of the Cybersecurity Strategy in the organization and the business must lead with driving Cybersecurity in the organization.

Based on the above, the key question in finalizing a Cybersecurity Organizational Structure is whether Cyber should report through Chief Information Officer (CIO) or as a separate Governance, Risk & Compliance (GRC) or Risk Management function directly to the CEO. It is recommended that the Chief Information Security Officer (CISO) function be established and have reporting direct to the CEO instead of reporting through the CIO. One reason for supporting this structure is the fact that Technology and Cyber have different mandates, priorities, and objectives. Technology individuals usually are in a hurry to meet deadlines for enabling the business while Cyber often insists on implementing the project cybersecurity controls, even if that means some delays. Another important reason is to provide segregation of duties. Cyber is often auditing the work of Technology and having them in the same reporting chain can make Compliance an issue. Finally, in the automotive sector, Cyber has a much more wide-ranging definition and is not strictly limited to the Information Technology domain.

Below are the key principles for establishing a Cybersecurity Structure in the Automotive Organization

- Establish the Chief Information Security Officer (CISO) role and assign ownership & accountability of the Cyber Governance Program to that individual
- Create a Cyber Innovation Steering Committee (CISC) that defines the relationship between CISO and other C-level executives such as CEO/COO/CFO/CIO and enables input from all senior executives for the Cybersecurity initiatives in the organization
- Define Governance, Risk & Compliance roles under the CISO that evaluates the risk appetite of the organization and enable a Governance Framework for the organization
- Define Cybersecurity Operations role under the CISO that actively manages the threats to the organization on a continuous basis
- Create Cybersecurity Training & Awareness roles under the CISO for ensuring a culture change with regards to Cybersecurity in the organization
- Establish a separate role under the CISO for Cyber Supply Chain Risk Management (C-SCRM) for ensuring compliance of vendors to the Cyber Governance Program of the organization
- Ensure that HR is fully involved in enabling the Cybersecurity Organizational Structure and has talent management in this field as a top priority

Below is the recommended Organization structure for medium-sized organization

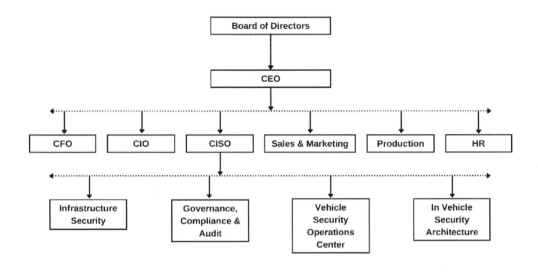

Figure 15 Organization Structure for Cyber Ownership

Cyber Asset Management

Asset Management is such a critical aspect of Cybersecurity that most Cyber Governance standards and frameworks have it as its first or, maximum, second category. A good example of this is the NIST Cyber Security Framework[4] which has Asset Management as its first category. The reason for this is obvious. If your organization does not know what assets it owns and which asset is essential for its continued operations, how can specific controls be put in place to protect its critical assets.

For an Automotive Manufacturing organization, there are many critical assets to consider:

---

[4] https://www.nist.gov/cyberframework

- In today's connected world, the crown jewels of any organization is the Data itself. This data may include Intellectual Property (IP), design and build processes, customer lists, employee details, business strategy, and marketing & branding plans.
- Another important asset is the infrastructure of the organization. This would include the IT network, IT hardware, OT components, Robotics, Operating Systems, applications, and the Cloud implementations.
- Physical Locations of the organization might also be considered an asset. Such an asset could be the assembly plant, the office building, or the data center where the data resides.
- Supplier Relationships should be considered a critical asset for any manufacturing organization. This is especially true in a global supply chain which depends on a borderless world for delivery of critical components just in time for completion of the final product.
- Another especially important Cyber asset are the people of the organization. This asset is often overlooked but it should always be remembered that people are the weakest link when it comes to cyber.

Cyber Risk Management

In today's connected world, Cyber Risk Management is an important component of managing the overall risk to the organization. This has become increasingly obvious as organizations become more reliant on technology for their continued operations and business success. Despite some investment on IT Security technology, organizations continue to suffer significant financial & reputational losses to their business due to Cyber breaches & technology interruptions. This means that Cyber has become a top business issue and Boards & Executive management are increasingly concerned about Cyber risk and want to better manage it. As a result, Cyber Risk Management of the whole organization has become an integral part of the overall Risk Management strategy and is used to underpin any Cyber initiatives launched in the organization.

Legal, Regulatory and Contractual Obligations

As our society becomes more dependent on technology, Cyber is becoming a fundamental part of the legal, regulatory & contractual framework of any manufacturing organization. An organization must adhere to Cyber specific for any & every jurisdiction it operates in. Many jurisdictions mandate Cybersecurity and are now even penalizing organizations for data breaches if PII (Personally Identifiable Information) is impacted. Such laws could be at the country-level or on Province/Stage level.

Industry regulators as well as contractual obligations is another area where Cyber obligations need to be highlighted and must be complied with. For the large-scale roll-out of Connected & Autonomous Vehicles (CAVs) on our roads, one major impediment is the challenge of finalizing the regulations applicable forCyber Governance. Once this is finalized, it would place even more importance on Governance, Risk & Compliance for meeting Cyber specific fiduciary obligations.

Cybersecurity Management Frameworks

Due Care & Due Diligence must be undertaken to ensure that all critical assets of an organization are protected. This includes the ability of the Executive Management to ensure that a Cyber Risk Assessment has been carried out in the organization and specific controls implemented to mitigate risks.

Cybersecurity Management Frameworks should be implemented to ensure that Cyber best practices are followed in the organization. There are many Cyber standards, frameworks, regulations & best practice guidance that may be applicable to an automotive manufacturing organization, including:

- ISO 21434 Road Vehicles Cybersecurity Engineering Standard
- NIST Cyber Security Framework (CSF)
- ENISA Good Practices for security of Smart Cars
- UNECE WP.29 R155 & R156 Cybersecurity Regulations
- Transport Canada Vehicle Cybersecurity Guidance
- Trusted Information Security Assessment Exchange (TISAX)
- General Data Protection Regulation (GDPR)
- Payment Card Industry Data Security Standard (PCI DSS)
- Sarbanes-Oxley (SOX) 404
- ISO 27001 Information Security Management System

Many of the above frameworks are further elaborated upon in chapter 8.

Cybersecurity Controls

The Cybersecurity Controls are specific measures that an Automotive Organization can implement to prevent, detect, and mitigate to security threats so that Cyber risks are considerably reduced (residual risks). A business can manage & mitigate Cyber threats by deploying these measures and thus, safeguard their crown jewels.

Although, every organization must review Cyber frameworks and implement the controls that are relevant to its environment, below are some of the most important controls:

- Identity Management & Access Control
- Security Architecture
- Perimeter Security
- Data Protection
- Cloud Security
- Encryption
- Physical Security
- Incident response Plan
- Vulnerability & Patch Management
- Anti-virus Solutions
- Employee Cybersecurity Awareness & Training
- Device (Vehicle) Management
- Service Provider Assessments
- Operations Security

Business Continuity and Incident Management

In the event of a major disruption or emergency, it is critical for an organization to have a defined Business Continuity Framework and an Incident Management Plan. Both together help to ensure that proper planning is in place for the management of any eventuality. They also provides framework for a coordinated response in the event of a Cyber breach. Finally, they also established the roles & responsibilities for various activities for response, recovery and business continuity and enable training for these specific roles & responsibilities.

An organization's Business Continuity Framework (BCF) will have the following core objectives:

- To minimize the impact of any eventuality on the organization's business operations and services
- To define the ownership of the corporate business continuity planning structure and process
- To establish a culture throughout the organization that enables the business continuity planning
- To identify the critical business processes and business functions of the organization
- To ensure that employees are trained in business continuity planning
- To ensure that communication to both internal & external stakeholders is properly managed during an emergency
- To finalize emergency response, business resumption and disaster recovery plans
- To regularly review the BCF and conduct periodic testing of the BCP & DRP to ensure readiness of the organization to face any unforeseen eventuality

Employee Cyber Awareness Training Program

In Cyber Governance, it is a common saying that the human being is the weakest link. The most stringent technical controls cannot prevent a human being from giving his or her password to a malicious individual impersonating to enable social engineering. Therefore, it is critical for an organization to ensure that the Cyber Culture is changed in the organization through regular and continuous Cybersecurity Education & Training carried out for ALL employees of an organization.

This training program should follow a few key principles

- It should incentivize and reward good Cyber behavior in employees
- It should be engaging and fun
- It should be communicated through multiple channels and formats
- It should be done on an ongoing and continuous basis

Vulnerability Assessment & Patch Management

Vulnerability Assessment is a methodology employed to identify the systems connected to an organization's network or used for delivering its services from the Cloud. This methodology would also apply to the systems in the CAV itself. Once the systems in any vehicle (or organizational network) are identified, these systems are then scanned to identify the Operating systems, applications, ports, and services being used on that system. Finally, any weaknesses in the software installed on these connected systems are also highlighted during this process.

Patch management is a pro-active approach to managing the weaknesses or vulnerabilities identified during the Vulnerability Assessment process. This includes

- Verifying & Validating vulnerabilities to classify the severity level of the vulnerability identified and the risk it presents to the organization if it is exploited
- Mitigating vulnerabilities is the analysis of preventing the specific vulnerability from being exploited if no patch is available or cannot be immediately implemented for various reasons such as application breakdown. Examples of such mitigation is taking an affected system offline or any other possible workarounds.
- Patching vulnerabilities is the final phase of this process and involves getting patches from the vendors of the affected systems and applying them to the identified systems in a timely way. Depending on the system, it can sometimes be an automated process. However, more frequently, before any patching is done on a Production system, patch testing is carried out in a Staging environment. This is a time consuming and complex process that involves rigorous testing of business applications but ensures that all systems are properly operating in the Production environment after the patch has been applied.

Cyber Supply Chain Risk Management Program

A critical element of an organization's Cyber profile is to ensure that all their third-party service providers also abide by Cyber best practices. In today's connected world, this is an important element for the overall Cyber profile of the organization, as an inter-connected eco-system is only as secure as its least secure member. In the automotive manufacturing supply-chain, this interdependence on external stakeholders becomes even more amplified due to the OEM's reliance on the supply chain to thrive and succeed in the industry.

Developing a holistic and robust Cyber Supply Chain Risk Management Program is thus an intrinsic obligation of any OEM, Tier 1, Tier 2 and other businesses operating in the Automotive Supply Chain. No single entity can ensure the safety and security of the end-product if collaboration and trust has not been established through a comprehensive Cyber Supply Chain Risk Management Program. In addition to ensuring assurance in a supplier's Cyber program, this ongoing engagement has many additional benefits. This is because it builds the processes, communication and relationships between various stakeholder entities that may be needed during a major Cyber incident for incident mitigation and leads to a closer business relationship between the entities in the eco-system.

Measuring an Effective Cyber Governance Program

A critically important but often overlooked element of a Cyber Governance Program is the measurement of its effectiveness. It takes great effort for an organization to implement a Cyber Framework to protect its "Crown Jewels" but the real question is whether this effort is successful? Not only is this measurement critical for the effectiveness of your cyber protections, but it is also essential for the Board & the Executive Management for informed decision making to ensure continued business success. However, as mentioned, this aspect of Cyber Governance Program has been severely lacking. An example of this is found in the 22nd Annual Global CEO Survey[5] carried out by PWC, a global consultancy, in 2019. It found that only 22% of chief executives believed that the risk exposure data they receive as part of their Governance Program was comprehensive enough to inform their decisions. An astounding and alarming part of this survey was that this figure i.e., 22% of CEOs who found insufficient risk data to inform their decisions, remained unchanged for 10 years.

---

[5] https://www.pwc.com/gx/en/ceo-survey/2019/report/pwc-22nd-annual-global-ceo-survey.pdf#page=29

Establishing effective Cyber Governance KPIs

The key word in establishing effective Cyber KPIs is "effective". The measurement of the effectiveness of an organization's Cyber Governance is only as good as the metrics that are being measured. To help develop effective Key Performance Indicators (KPIs) and to ensure risk management teams align with the business goals, the Internet Security Forum (ISF)[6] has developed a four-phase, practical approach to developing KPIs and KRIs. ISF has designed this approach to be utilized at all levels of an organization and consists of four phases:

- Establish relevance by engaging to understand the business context, identify common interests and develop combinations of KPIs and Key Risk Indicators (KRIs)
- Generate insights by engaging to produce, calibrate and interpret KPI/KRI combinations
- Create impact by engaging to make recommendations and make decisions about possible actions that can be subsequently undertaken
- Learn and improve by engaging to develop learning and improvement plans

The main idea at the core of ISF's methodology is to ensure the engagement of all stakeholders that are concerned with an organization's cybersecurity. This approach which fosters "Cyber Engagement", builds relationships and improves understanding, allowing the cybersecurity and risk management functions to better respond to the needs of the business.

Cyber Governance KPIs to measure

Although most organizations have different perspectives on the Cyber Governance KPIs that need to be measured, below are a few of the most important ones that should be measured:

- **Cyber ownership structure of the organization:** Is there ownership of the various Cyber roles & responsibilities and are these properly defined?

---

[6] https://www.securityforum.org/news/how-cisos-can-create-security-kpis-and-kris-25-06-2015/

- **Cyber SME Skill Management:** Does the organization have a comprehensive talent management program to ensure qualified Cyber SMEs are available to deliver on the Cyber responsibilities?
- **Adherence to Industry Standards & Frameworks:** Is the organization aware of the contractual and regulatory obligations that it needs to meet to ensure continued success in business operations?
- **Access management:** Are all users in the organization identified and whether all users with administrative access are recognized as having enhanced access privileges?
- **Cyber breaches & Intrusion attempts:** Is there a clear understanding of the total intrusion attempts that have been carried out against the organization's data and infrastructure assets? Has there been successful breaches to the organization's network and if yes, which attack vectors were used?
- **Device readiness:** How long does it take to have a new patch installed on any device/product/vehicle and how many devices/products/vehicles are fully patched and up-to-date?
- **Incident Management:** Measurement of three key KPIs that ensure that ensure Incident Management significantly improves in the Organization. These are 1) Mean Time to Detect (MTTD), 2) Mean Time to Contain (MTTC) and 3) Mean Time to Resolve (MTTR)
- **Cybersecurity awareness training results:** Identification of whether all employees have undergone Cybersecurity awareness training and whether it helped improve the overall Cyber understanding of these individuals.
- **Cloud Security & SaaS apps:** The organization should have clear identification of the apps that are being migrated to the Cloud and their interdependencies. Also, it is imperative that the third-party SaaS apps which the organization is utilizing be known and identified.
- **BCP & DRP:** Are the Business Continuity & Disaster Recovery Plans in place? Has there been any significant and periodic testing? Are there any significant improvements that need to be made in the organization's BCP & DRP?
- **Alignment with Comparable Industry Entities:** Effort should be made to ensure metrics are in place which can give companies guidance on

comparable Cyber Governance efforts and initiatives with the wider automotive industry.

What you Learned in this Chapter

This chapter has given a holistic overview of Cyber Governance for an automotive organization. Areas covered in this chapter include:

- The criticality of Corporate Governance implemented in any organization.
- The need for including a holistic Cyber Governance program in the Automotive Organization's overall Corporate Governance structure.
- The building blocks of a comprehensive Cyber Governance Program. Among other building blocks, this includes defining the Cyber Strategy, the organizational Cybersecurity structure, the Cybersecurity Controls deployed, the Cyber Supply Chain Risk Management program and many other elements.
- The importance of measuring the effectiveness of the Cyber Governance program and defining KPIs and KRIs to measure this effectiveness.

Chapter 7: Defining and Measuring Cyber Risk

Chapter Overview

This chapter delves into the concepts of Risk Management. It begins with expounding the methodologies of identification of Cyber Risk, including Gap Analysis and TARA. It then goes on to define the various terms associated with Cyber risk. An important component of this chapter are the risk management strategies which explain how to manage and mitigate risks. It then covers the types of risk analysis and the criticality of selecting appropriate safeguards/controls. Finally, the chapter gives an overview of the various Cyber risk assessment methodologies.

Identifying Cyber Risk

One of the most vital tasks of the Chief Information Security Officer (CISO) is to ensure that Threat Analysis & Risk Assessment (TARA) exercises are carried out on a regular basis in the organization. These are also sometimes called Threat & Risk Assessments (TRAs) or simply Risk Assessments. These are done to enable a strategic approach to Risk Management and answer the following questions:

- What are the assets at risk that are owned by the organization?
- What is the value at risk, as associated with the identity of information assets?
- What are the threats that could impact the organization and what is the annualized financial consequences of these threats?
- Has any Risk mitigation analysis been carried out using a formal methodology?
- What are the costs for Risk mitigation and are there any alternative options?

The result of such a strategic approach to risk management is to ensure that the organization's security controls are defined based on a detailed analysis of the Cyber threats to the organization as opposed to trying to only implement industry best practices. The aim here is to make the controls specific to the organization itself and ensure that the Cyber profile of the organization is considerably strengthened.

There are many ways to perform risk assessments, starting from simple gap analysis to more detailed TARA methodologies. However, all Cyber Risk Assessment methodologies follow the same basic approach. They all start with identifying the organization's critical assets and then evaluating the threats to those assets based on any weaknesses or vulnerabilities that can be exploited in those assets. If a specific vulnerability exists in the organization's infrastructure that can be targeted by a malicious attacker through a specific threat, there would be risk impact to the organization. Specific safeguards or controls would then have to be implemented to reduce the risk to a more acceptable level, residual risk, to the organization.

Gap Analysis

Gap Analysis is done at the 30,000 feet level and provides a comparison of the organization's security program with Cybersecurity best practices. A Gap Analysis is usually done against a specific Governance Framework such as NIST CSF, ISO21434, ISO27001/2, GDPR, PIPEDA or others. However, the steps required to carry out an effective Gap Analysis remain the same. These include:

- Selecting the appropriate Governance Framework to carry out the Gap Analysis
- Evaluating the Policies, Processes & Procedures in the Organization against the best practices of the chosen Governance Framework
- Evaluating the technology used in the Organization against the controls defined in the Governance framework
- Analyzing the data and develop a roadmap for short-, med- and long-term enhancement of Cyber Security in the Organization

Threat Analysis & Risk Assessment

A Threat Analysis & Risk Assessment (TARA) is more detailed than a Gap Analysis and follows defined methodologies to explore the Risk Impact to each asset owned by the organization. Although most commonly, a Qualitative approach to risk assessment is used, many Cyber-aware organizations carry out Quantitative risk assessment. Such a Quantitative risk assessment approach aims to come up with a monetary value for the impact of various threats to your organization. This enables a more informed decision-making by the executive management and allows the organization to develop an appropriate strategy for handling the various highlighted risks.

All formal methodologies that are used to carry out a TARA are focused on one main objective: to evaluate the immense number of possible threats to an organization and ultimately, distill these into a manageable number of scenarios that are most likely to occur. This distillation of all possible threats is done based upon the objectives, methods, capabilities, and desires of malicious actors who might wish to harm the organization. These TARA methodologies also incorporate the existing controls in place at the organization and provides a roadmap for new initiatives for risk mitigation.

Cyber Risk Management Terms and Definitions

Below are some of the key definitions for Cyber Risk Management

Assets

An asset can refer to any data, resource, process, product, computing, or physical infrastructure that is owned by an organization. Usually, the most critical assets are highlighted during the risk assessment process, and these can be considered the "Crown Jewels" of the organization.

Threats

A threat is a broad category of events whose occurrence can cause an undesirable impact on the organization. An example of this is "Hacking".

Threat Agent

A threat agent is a very specific threat whose occurrence can cause an undesirable impact on the organization. For example, the "Hacking" threat category may be broken down into specific Threat Agents such as "Wannabees", "Script Kiddies", "Fully Capable", "Elite Hackers" & "State Actors".

Vulnerabilities

A vulnerability is a weakness in an asset owned by the organization that can be exploited by a specific threat or threat agent.

Safeguards or Controls

A safeguard or control is a countermeasure employed by the organization to ensure that the risk associated with a specific threat is considerably reduced to a manageable level.

Cyber Risk

The Oxford learners dictionary defines Risk[1] as "the possibility of something bad happening at some time in the future". Thus, risk is inherent in any situation that could be dangerous or have an adverse impact to the organization under specific circumstances in the future.

Although, there is a broad range of risks that can affect an organization, Cyber risks are becoming more and more critical for the smooth functioning and indeed, the business success of any organization. In today's data driven world, Cyber risks may also significantly impact the brand of an organization and lead to significant loss in the reputation of the organization as a credible business partner. Specifically, for the Automotive manufacturing sector with its integrated supply chain, mitigating Cyber risk has become essential for the business survival of OEMs, Tier 1 & Tier 2 Suppliers. This is due to several factors including consumer demand, Industry 4.0, Connected car technology, OT & IT integration, and more stringent regulatory frameworks.

---

[1] https://www.oxfordlearnersdictionaries.com/definition/english/risk_1

Cyber Risk is the probability of exposure or loss resulting from the occurrence of any event related to data, technical infrastructure and use of technology in the organization. This event may be a hacking attack or data breach on your organization but it may be something more mundane such as system misconfiguration or power loss that brings down major technology assets.

To define Risk to an organization, the following must be defined:

- The actual threat
- The possible consequences of the threat
- The probable frequency of the occurrence of a threat
- The extent of how confident the organizations are that the threat will happen

Any identified risks must be listed in a "Risk Register" maintained by the Risk Management team of the organization. The information in the "Risk Register" must be continuously updated to ensure that the organization is cognizant of the risks that may impact its business.

Risk Management Strategies

Once Risk levels have been determined by carrying out a TARA, an organization can use several different risk management strategies. These include:

- Risk Reduction
  - Taking measures to mitigate risk from specific threats. Common examples of such an approach include putting up a firewall to ensure malicious traffic is not allowed into the organization's network or using Two-Factor Authentication to ensure more stringent authentication to the organization's systems. However, in the Automotive sector, a detailed analysis must be done to ensure effective Risk Reduction measures are carried out for the domain of automotive cybersecurity.

- Risk Transference
    - Assigning or transferring the potential cost of a loss to another party. The best example of this is to have Cyber Insurance to ensure that the cost of any potential undesirable event is born by another party. However, insurance companies are still evaluating the Cyber risk to the vehicle itself and have yet to develop effective Cyber Insurance policies for the Connected & Autonomous Vehicles (CAVs).
- Risk Acceptance
    - Accepting the level of loss that will occur. This is usually not the best strategy but is often employed as Cyber is considered additional cost by the executive management. Usually, this strategy leads to the build-up of "Security debt" which manifests itself in cyber breaches and loss of business operations due to disruption in technology assets.
- Risk Avoidance
    - Not performing any activity that may carry risk. An example of this is to not procure any technology that is deemed to have "high risk" and use alternate means to achieve the same business function. With the advent of ever greater technologies being implemented in CAVs, this is really not an option for an automotive organization to survive and thrive as we move into a more connected future.

Monetizing Cyber Risk

Recognizing the financial impact of Cyber risk is a high priority item for both the Executive Management as well as the Board of Directors of any organization. It is thus essential to carry out Risk Assessment (RA) exercises that evaluate possible cyber events such as hacking and provide estimates of the effect to the organization related to specific cyber incidents. This would enable the Executive Management to take decisions to mitigate the risk to the entity based on informed analysis.

There are two major types of risk assessments: Qualitative Risk Analysis and Quantitative Risk Analysis.

Qualitative Risk Analysis

A qualitative risk analysis does not provide a monetary value at the end of process. However, the aim of a qualitative risk analysis is to provide the risks faced by the organization as a function of both the likelihood of a specific risk event occurring and the impact it will have on the targeted asset should it happen. Both the likelihood and the impact are measured in the form of High, Medium, Low and a risk matrix is used to determine severity of the risk to the organization. Based on the risk level, the executive management can then take decisions on control implementation for risk mitigation.

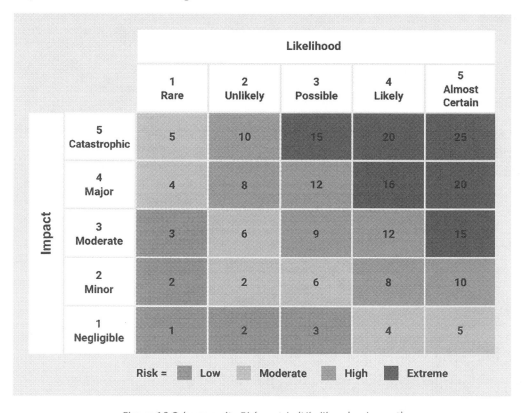

Figure 16 Cybersecurity Risk matrix (Likelihood vs Impact)

A qualitative risk analysis is generally done by practitioners and entities who do not have the expertise to carry out a risk analysis based on monetary values or do not have the resources to carry out a more detailed risk analysis.

Below is the methodology for qualitative risk analysis:

- Scenario is written for specific threat
- Scenario is reviewed for likelihood and impact to the organization if that specific threat occurs
- RA defines the level of risk from the specific threat to the organization as high, medium, or low depending on likelihood and impact to the organization
- RA team evaluates and controls the various safeguards/controls that can further reduce the risk
- RA team walks through each finalized scenario and defines the level of residual risk to the organization after safeguard is implemented
- Final report is submitted to the Executive Management

Quantitative Risk Analysis

The aim of a quantitative risk analysis is to assign independently objective monetary value to the components of the risk assessment and to the assessment of potential losses. This process begins with the risk management team estimating the potential losses to assets by determining their value. They go on to analyze potential threats to the assets and then define the financial loss that would occur if a specific asset suffered a disruption from a specific threat.

There are some key definitions for carrying out quantitative risk analysis. These are:

- Asset Value (AV)   = Total Value of any Asset
- Exposure Factor (EF)   = Percentage (%) of asset loss caused by threat
- Single Loss Expectancy (SLE)   = AV x EF
- Annualized Rate of Occurrence (ARO)   = Frequency of threat occurrence per year
- Annualized Loss Expectancy (ALE)   = SLE x ARO

The following scenario clarifies the process used for quantitative risk analysis. Let us consider an automotive organization that has a critical system used in the development of a CAV and the Asset Value of this critical system has been estimated at $150,000. It is estimated that if that system is disrupted by a Cyber breach (Threat A), the entity will suffer a one-time loss of 25% of the system AV. Also, it is estimated that such an undesirable event may happen once in ten years i.e., an ARO of 0.1. This leads to the below calculations:

Asset Value		=	$150,000.00
SLE	= $150,000 x 0.25	=	$37,500.00
ALE	= $37,500 x 0.1	=	$3,750.00

The above calculations imply that the entity can implement controls or safeguards worth $3,750.00 or less per year to secure this specific asset against this specific threat (Threat A).

It is important to note there that this amount of $3,750.00 is only for the control which protects the asset from Threat A. There might be another greater threat (Threat B) to the same asset which might have a higher ALE and the controls defined for Threat B might also protect against risks from Threat A.

Safeguard/Control Selection

The main aim of carrying out a detailed risk analysis is to provide the Executive Management and the Board of Directors enough information to define the Risk Management strategy. If a "Risk Reduction" strategy is selected, then safeguards or controls need to be implemented that would reduce the amount of risk to the organization. This final level of risk is also called Residual Risk and the aim of implementing controls should be to have the residual risk for specific threats at an acceptable level.

It is important to highlight that this safeguard selection exercise must be carried out for all assets for all relevant threats. However, it might be the case that the one safeguard might be able to reduce the risk from multiple threats. An example of this is the implementation of a firewall at the network perimeter which would not only protect against intrusion attempts from outside the firewall but could also ensure that only permitted internal traffic is able to be transmitted outside the organization.

Asset Valuation Process

An asset valuation process is critical for the successful completion of the risk analysis process. It is essential for many reasons including for Cost / Benefit Analysis (CBA) and for determining insurance costs. It also facilitates in the safeguard selection decision-making process and finally, it may also be required for legal obligations and to show "due care" was done.

Safeguard Cost / benefit Analysis

Safeguard selection cannot be completed without carrying out a detailed Cost/Benefit Analysis (CBA). Such an analysis should include the following input:

- The purchase, development and /or licensing costs of the safeguard
- The physical installation costs and disruption to normal production during installation and testing of the safeguard
- Normal operating costs, resource allocations and maintenance / repair costs

It is possible to calculate the value of the safeguard to the organization through this CBA exercise. The value of the safeguard to the organization can be defined as

Value of the safeguard to the organization

= ALE before Safeguard – ALE after Safeguard – annual Safeguard cost

Safeguard Recovery Ability

Any safeguard selection must be able to effectively recover from a disruptive event. This means that during and after a reset condition, the safeguard must provide the following:

- No asset destruction during activation or reset
- No covert channel access to or through the control during reset
- No security loss or increase in exposure after activation or reset
- Defaults to a state that does not give any operator access or rights until the controls are fully operational

Cyber Risk Management Program Core tasks

Cyber Risk Management is a complex and ongoing activity in any major entity. The cyber risks to an organization are continuously changing in the complex connected world of today. Unforeseen events such as COVID19 pandemic and the resultant Work From Home (WFH) environment provide unnecessary Cyber exposure to an organization if it is not prepared for such unforeseen circumstances. Therefore, it is essential for an organization to have a comprehensive Risk Management Program to effectively mitigate the impact of any adverse circumstances.

The below list provides the core tasks that are inherent in a formal Cyber Risk Management Program

- Define the Cyber Risk Management Policy
- Establish & Fund a Cyber RM Team
- Establish & Approve a defined Cyber RM Methodology & Tools to be used for this purpose in the organization
- Identify & Measure Cyber Risk on an ongoing basis
- Project Sizing
- Scope of the Information Protection Environment
- Asset Identification & Valuation
- Vulnerability Analysis
- Threat Analysis
- Cyber Risk Evaluation
- Interim Reports & Recommendations

- Information Protection Environment
- Establish Cyber Risk Acceptance Criteria
- Safeguard Selection and Risk Mitigation Criteria
- Cost/Benefit Analysis
- Implementation of selected Controls to mitigate Risk
- Calculate Residual Risk
- Final Report

Cyber Risk Assessment Methodologies

There are many Risk Assessment methodologies which various cyber-focused entities, frameworks and national governments have formulated. Some of these are:

- NIST RMF
- Harmonized TRA (HTRA)
- Austrian IT Security Handbook
- Cramm
- Dutch A&K Analysis
- Ebios
- ISAMM
- ISF Methods
- ISO/IEC 13335-2
- ISO/IEC 17799
- ISO/IEC 27001
- IT-Grundschutz
- Magerit
- Marion
- Mehari
- MIGRA
- OCTAVE
- RiskSafe Assessment

The major Cyber Risk Assessment methodologies are highlighted below. However, a Risk Management Methodology must be adopted by an organization based on its jurisdiction, industry, SME expertise and other related criteria to be able to achieve effective Risk Management.

NIST RMF

NIST Risk Management Framework (RMF)[2] is composed of comprehensive risk management guidelines issued by National Institute of Standards & Technology in the United States of America. As per NIST, "The Risk Management Framework provides a process that integrates security and risk management activities into the system development life cycle. The risk-based approach to security control selection and specification considers effectiveness, efficiency, and constraints due to applicable laws, directives, Executive Orders, policies, standards, or regulations."

The NIST RMF is composed of six steps:

1. Categorize
2. Select
3. Implement
4. Assess
5. Authorize
6. Monitor

---

[2] https://csrc.nist.gov/projects/risk-management/rmf-overview

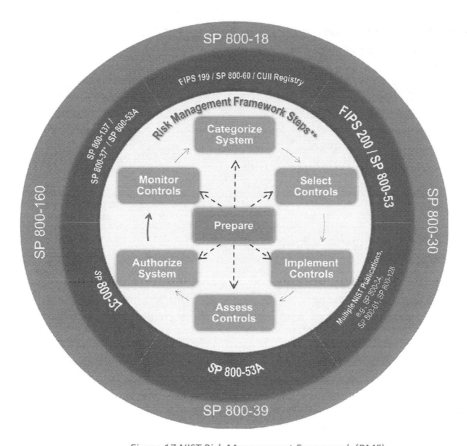

Figure 17 NIST Risk Management Framework (RMF)

As outlined in the above figure, there are several NIST special publications that can help support a practitioner to better understand the entire NIST RMF methodology as formulated by NIST.

Harmonized TRA

Harmonized Threat & Risk Assessment (HRTA)[3] is a Canadian government approved methodology for assessing the impact of risk to the organization. This is a methodology issued as an unclassified publication, under the authority of the

---

[3] https://cyber.gc.ca/en/guidance/harmonized-tra-methodology-tra-1

Chief, Communications Security Establishment (CSE) and the Commissioner, Royal Canadian Mounted Police (RCMP).

HTRA has the following steps:

1. Identify Assets
2. Identify Threats
3. Identify Vulnerabilities
4. Calculate Residual Risks
5. Monitor & Report

Figure 18 Harmonized TRA

OCTAVE

OCTAVE[4] is the risk assessment methodology developed by the Carnegie Mellon University in USA and has been widely adopted including by ENISA[5], the European Union Agency for Cybersecurity.

---

4 https://resources.sei.cmu.edu/library/asset-view.cfm?assetid=13473
5 https://www.enisa.europa.eu/topics/threat-risk-management/risk-management/current-risk/risk-management-inventory/rm-ra-methods

OCTAVE stands for Operationally Critical Threat, Asset, and Vulnerability Evaluation. As per OCTAVE, "it defines a comprehensive evaluation method that allows an organization to identify the information assets that are important to the mission of the organization, the threats to those assets, and the vulnerabilities that may expose those assets to the threats. By putting together, the information assets, threats, and vulnerabilities, the organization can begin to understand what information is at risk. With this understanding, the organization can design and implement a protection strategy to reduce the overall risk exposure of its information assets."

ISO 21434 TARA Methodology

ISO 21434 has detailed a comprehensive TARA methodology specifically applicable for the development of CAVs. Clause 15 of this standard defines the objectives for TARA as

a) identify assets, their cybersecurity properties and their damage scenarios;

b) identify threat scenarios;

c) determine the impact rating of damage scenarios;

d) identify the attack paths that realize threat scenarios;

e) determine the ease with which attack paths can be exploited;

f) determine the risk values of threat scenarios; and

g) select appropriate risk treatment options for threat scenarios.

A key area of enabling the ISO 21434 TARA modeling is understanding the damage scenarios as well as the Threat Catalog applicable to the CAV. This is where the knowledge, expertise and experience of an Automotive Cyber SME, as opposed to an IT Cyber SME, would be critical to effectively completing the TARA excersize.

Threat Modeling

Threat Modeling is an important component of the risk assessment process. Threat modeling helps the security practitioner to comprehend the threats to a specific asset and is an input to the risk assessment process to ensure that specific security controls are implemented to reduce the risk. Threat Modeling exercises can also be used for Cyber Wargaming to explore different scenarios of possible malicious actor attacks and countermeasures against those security breaches.

There are several different methodologies defined for Threat Modeling. The most well known of these are:

- STRIDE
- DREAD
- PASTA
- VAST
- Trike
- OCTAVE
- NIST

In this book, STRIDE, DREAD and PASTA will be defined in more detail in this section as these are the most used Threat Modeling methodologies. NIST and OCTAVE have already been defined in the previous section and should be considered as more holistic risk assessment frameworks.

STRIDE

STRIDE was first developed at Microsoft in the late '90s and has been used extensively. STRIDE stands for six categories of possible threats and these threats are based on violation of a specific property of the CIA triad.

These threats are:

- **Spoofing**
  - Spoofing means impersonating another person or computer and this violates the *integrity* principle of the CIA Triad as the authenticity of the interaction is invalidated.

114

- **Tampering**
  - Tampering of or with the data violates the *Integrity* principle of the CIA Triad.
- **Repudiation,**
  - Repudiation means the denial of an action and *non-repudiation* ensures that the action is linked to the individual who performed it. This would mean that repudiation cannot be performed. Repudiation violates *all three principles* of the CIA Triad.
- **Information disclosure**
  - Information disclosure is the reverse of the *Confidentiality* principle of the CIA Triad.
- **Denial of service**
  - Denial of Service is the reverse of the *Availability* principle of the CIA Triad.
- **Elevation of privilege**
  - **Elevation of privilege** violates *authorization* which can impact *all three principles* of the CIA Triad.

DREAD

DREAD was conceived of as an add-on to the STRIDE model that allows modelers to rank threats once they have been identified. DREAD has five categories of questions which can be answered to define the severity of the threat. These five categories together make up the DREAD acronym.

- **Damage potential**
  - The potential of damage from the attack should be enumerated if a specific vulnerability is exploited?
- **Reproducibility**
  - If it is easy to reproduce the attack, then this means that there are greater potential for recurring attacks of this nature.
- **Exploitability**
  - The ease with which the attack can be launched implies less expertise is required by the malicious actors. Thus, it can be determined whether script kiddies or more expert hackers can launch these attacks.

- **Affected users**
  - o The percentage of users in the organization which can be affected by the attack and the level of organization business interruption that can happen.
- **Discoverability**
  - o How easy is it to find the vulnerability? If more effort is required to find the specific vulnerability to exploit, then it would be harder for an attacker to use this vulnerability to enable a security breach.

Each of these questions is answered with a rating between one and three. The total sum of the score of all five categories can then be used to prioritize specific vulnerabilities to safeguard.

PASTA

PASTA stands for *Process for Attack Simulation and Threat Analysis.* PASTA is a seven-step process and aims to align the technical security requirements with business objectives of the organization. However, the whole process is complex as each step consists of several sub-steps. The overall sequence of PASTA is as follows:

- Define objectives
- Define technical scope
- Application decomposition
- Threat analysis
- Vulnerability and weaknesses analysis
- Attack modeling
- Risk and impact analysis

The aim for PASTA was to have dynamic threat modeling and not to limit the threat modeling process to a moment in time. However, this needs considerable expertise to be developed in this field by the cybersecurity SME.

What you Learned in this Chapter

The chapter has provided the reader with the below aspects of Cyber Risk Management:

- A comprehensive overview of Cyber Risk Management concepts
- The importance of the Cyber Risk management to an Automotive Organization
- Details of various Cyber Risk Management strategies that any organization can utilize to mitigate Cyber risk
- The different Risk Analysis methodologies that can be used to determine the level of Cyber risk to the organization
- The Safeguard/Control Selection process
- The detailed list of the core tasks of the Cyber Risk program
- Formal Risk Management Methodologies
- Threat Modeling Methodologies

Chapter 8: Enabling Governance, Risk & Compliance (GRC)

Chapter Overview

This chapter articulates the Implementation of a Cyber Framework for an Automotive Organization to enable Governance, Risk & Compliance within it. The chapter begins by highlighting the criticality of Cyber Governance Frameworks towards enabling the basic functions of any compliance program in any organization and it goes on to describe the structure of Cybersecurity frameworks. Finally, the chapter goes on to enumerate the various Cyber frameworks that might be applicable in an Automotive Organization and describes the main frameworks in greater detail.

Cyber Governance Frameworks

Cyber Governance frameworks ensure that the basic functions for any compliance program of an information protection environment are met. This is done through establishing governance benchmarks through documentation describing Policies, Procedures, Processes, Standards and Guidelines applicable in the organization. These benchmarks can then be audited against to showcase compliance to the Cyber Governance framework. In the case of a Cyber breach, an established Cyber Governance framework also enables the organization to show that it has carried out "due care" and "due diligence" and help decrease the legal ramifications of the Cyber breach.

A well-designed and properly implemented Cyber Governance Framework ensures the following in any organization:

- Communicates the policies, standards, baselines, guidelines and procedure to all employees and other stakeholders of the organization

- Ensures that compliance to implemented Cyber policies in the organization is being carried out by all stakeholders to the organization's business success
- If compliance to the implemented Cyber policies of the organization is not being carried out, the Governance Framework should recommend the enforcement of the above through appropriate disciplinary measures
- Implements procedures for corrections in case of violations

Policies, Standards, Procedures, Baselines & Guidelines

It is important to understand the structure of a Cyber Governance Framework and implement the various elements that together form the entire Cyber Governance Framework. These are:

- **Policies**

  - Policies are high-level Cyber Governance documentation that regulate the use of data and the use of specific technology in an organization. Some examples of organization-wide cyber policies include Information Security Policy, Acceptable Use Policy, Incident Management Policy, Business Continuity Policy, Wifi Usage Policy, Asset Inventory Policy, Access Control Policy and Remote Work Policy. This is by no means an exhaustive list and a detailed list of policies is available in Appendix A of this book.

- **Standards**

  - Standards are detailed specifications that elaborate the usage of technologies in a uniform way in the organization. Some examples of Standards would be MS Windows Server Configuration Standard, MS Windows Workstation Configuration Standard, Linux Server Configuration Standard and Cryptography Standard.

- **Procedures**

  - Procedures are detailed steps that outline the methodology to perform a specific task. Examples of a procedure can be the patch management procedure or VPN configuration procedure.

- **Baselines**

  - Baselines are used to develop a consistent trend regarding the behavior of the organization's systems. A good example of a baseline is to monitor the traffic patterns on the network or the usage of an organization's web servers.

- **Guidelines**

  - Guidelines enumerate the Cyber best practices that are recommended for methodologies to use while securing systems, but these are not mandatory in the organization. For example, a recommendation might be to follow CIS benchmarks[1] to secure a system while installing it on the organization infrastructure.

---

[1] https://www.cisecurity.org/cis-benchmarks/

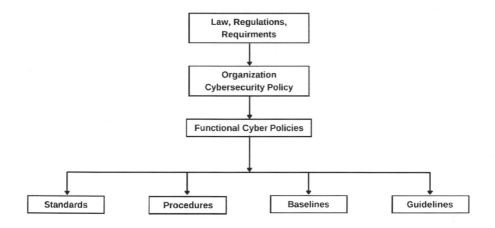

Figure 19 Cyber Governance structure

Organization Cybersecurity Policy

The Automotive Organization must have a high-level Organization Cybersecurity Policy that is applicable to the entire organization and related stakeholders. This policy is a general and high-level statement of policy from the Executive Management containing

- An acknowledgment of the importance of the data & related computing resources to the business success of the organization
- A statement of support for Cybersecurity throughout the enterprise
- A commitment to authorize and manage the definition of the lower-level policies, standards, procedures, baselines and guidelines i.e. the entire Cyber Governance Framework of the organization.

Types of Policies

The Cyber Governance Framework can define three different types of policies to implement the Governance Program in the organization. These are:

- Regulatory

These are the policies that ensure that an organization is following the standards in a specific industry. They also give the organization the confidence that they are following industry standards.

- Advisory

Advisory policies are not mandated to be followed but are strongly suggested.

- Informative

Informative Policies exist only to inform readers and are not mandated or strongly suggested.

All three types of policies work together to ensure the successful implementation and operation of the Cyber Governance program in the organization.

Cyber Standards & Frameworks

There are many Cyber Standards & Frameworks that can be adhered to by various organizations. The most relevant Cyber Standards & Frameworks for the Automotive sector are:

- ISO 21434 Road Vehicles Cybersecurity Engineering Standard
- NIST Cyber Security Framework (CSF)
- ENISA Good Practices for security of Smart Cars
- UNECE WP.29 R155 & R156 Cybersecurity Regulations
- Transport Canada Vehicle Cybersecurity Guidance
- Trusted Information Security Assessment Exchange (TISAX)
- General Data Protection Regulation (GDPR)
- Payment Card Industry Data Security Standard (PCI DSS)
- Sarbanes-Oxley (SOX) 404
- ISO 27000 Series (Information Security Management System)

ISO 21434 Road Vehicles Cybersecurity Engineering Standard

ISO21434: Road Vehicles -- Cybersecurity Engineering Standard has been jointly developed by the International Organization of Standardization (ISO) and the Society for Automotive Engineers (SAE). It was published at the end of August 2021 and can be used to develop a holistic Cybersecurity Management System (CSMS) in the automotive organization. The aim of this standard is to develop common terminology and criteria around key aspects of cybersecurity for the Automotive Sector. This standard helps identify the methodology which can be used to enable cyber controls in all aspects of vehicle development and operations. This means that by applying the controls available in the standard, companies will be able to demonstrate due care and due diligence related to cyber-threat prevention in vehicle development, operations, maintenance, and disposal.

Figure 20 ISO21434 Objectives

Purpose of ISO 21434

Below is the excerpt[2] from the ISO 21434 standard that elucidates the purpose of the standard:

"This document addresses the cybersecurity perspective in engineering of electrical and electronic (E/E) systems within road vehicles. By ensuring appropriate consideration of cybersecurity, this document aims to enable the engineering of E/E systems to keep up with changing technology and attack methods.

This document provides vocabulary, objectives, requirements, and guidelines as a foundation for common understanding throughout the supply chain. This enables organizations to:

- define cybersecurity policies and processes;

- manage cybersecurity risk; and

- foster a cybersecurity culture.

This document can be used to implement a cybersecurity management system including cybersecurity risk management in accordance with ISO 31000. This document is intended to supersede SAE J3061 recommended practice."

In addition, the below figure[3] shows the overall structure of the standard:

---

[2] ISO 21434 Road Vehicles Cybersecurity Engineering Standard
[3] ISO 21434 Road Vehicles Cybersecurity Engineering Standard

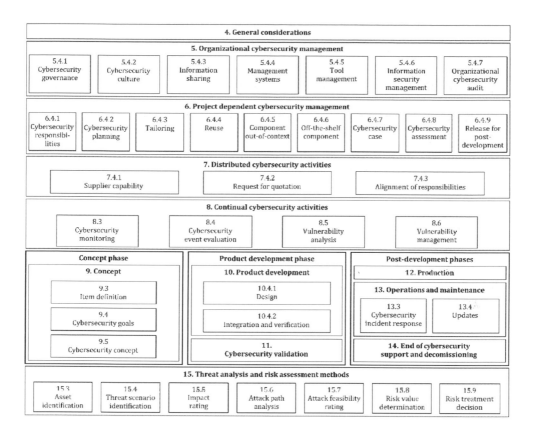

Figure 21 ISO 21434 Structure

Cyber Security Management System

The standard describes the critical need to develop and implement a Cyber Security Management System (CSMS) in the organization that showcases the management commitment to be compliant with the standard. The implementation of such a CSMS mandates several processes that must be in place in the organization as well as in the supply chain. Specific controls such as policies, procedures and technical controls must be designed to implement such a CSMS. However, the standard itself does not propose any specific technologies.

Importance of Threat Analysis & Risk Assessment (TARA)

An important aspect of the ISO 21434 Standard is the emphasis placed on Threat Analysis & Risk Assessment (TARA). The results of such a detailed assessment guide the activities in the product development stage of the vehicle. These results would also enable the adoption of Cyber Controls to safeguard the vehicle during its lifecycle. As is common with most cyber best practices, the standard emphasizes the need for a TARA, it does not describe a specific methodology for carrying out the TARA.

Work Products

Several Work Products (WP)[4] are specifically required by the standard to enable cybersecurity activities in the organization as part the implementation of this standard. These Work Products (WP) are enumerated below

---

[4] ISO 21434 Road Vehicles Cybersecurity Engineering Standard

Table 3 ISO 21434 Work Products

Sub-clauses	Work products
Organizational cybersecurity management	
5.4.1 Cybersecurity governance	[WP-05-01] Cybersecurity policy, rules and processes
5.4.2 Cybersecurity culture	[WP-05-01] Cybersecurity policy, rules and processes
	[WP-05-02] Evidence of competence management, awareness management and continuous improvement
5.4.3 Information sharing	[WP-05-01] Cybersecurity policy, rules and processes
5.4.4 Management systems	[WP-05-03] Evidence of the organization's management systems
5.4.5 Tool management	[WP-05-04] Evidence of tool management
5.4.6 Information security management	[WP-05-03] Evidence of the organization's management systems
5.4.7 Organizational cybersecurity audit	[WP-05-05] Organizational cybersecurity audit report
Project dependent cybersecurity management	
6.4.1 Cybersecurity responsibilities	[WP-06-01] Cybersecurity plan
6.4.2 Cybersecurity planning	[WP-06-01] Cybersecurity plan
6.4.3 Tailoring	[WP-06-01] Cybersecurity plan
6.4.4 Reuse	[WP-06-01] Cybersecurity plan
6.4.5 Component out-of-context	[WP-06-01] Cybersecurity plan
6.4.6 Off-the-shelf component	[WP-06-01] Cybersecurity plan
6.4.7 Cybersecurity case	[WP-06-02] Cybersecurity case
6.4.8 Cybersecurity assessment	[WP-06-03] Cybersecurity assessment report
6.4.9 Release for post-development	[WP-06-04] Release for post-development report
Distributed cybersecurity activities	
7.4.1 Supplier capability	None
7.4.2 Request for quotation	None
7.4.3 Alignment of responsibilities	[WP-07-01] Cybersecurity interface agreement

Sub-clauses	Work products
Continual cybersecurity activities	
8.3 Cybersecurity monitoring	[WP-08-01] Sources for cybersecurity information [WP-08-02] Triggers [WP-08-03] Cybersecurity events
8.4 Cybersecurity event evaluation	[WP-08-04] Weaknesses from cybersecurity events
8.5 Vulnerability analysis	[WP-08-05] Vulnerability analysis
8.6 Vulnerability management	[WP-08-06] Evidence of managed vulnerabilities
Concept phase	
9.3 Item definition	[WP-09-01] Item definition
9.4 Cybersecurity goals	[WP-09-02] TARA [WP-09-03] Cybersecurity goals [WP-09-04] Cybersecurity claims [WP-09-05] Verification report for cybersecurity goals
9.5 Cybersecurity concept	[WP-09-06] Cybersecurity concept [WP-09-07] Verification report of cybersecurity concept
Product development phase	
10.4.1 Design	[WP-10-01] Cybersecurity specifications [WP-10-02] Cybersecurity requirements for post- development [WP-10-03] Documentation of the modelling, design, or programming languages and coding guidelines [WP-10-04] Verification report for the cybersecurity specifications [WP-10-05] Weaknesses found during product development
10.4.2 Integration and verification	[WP-10-05] Weaknesses found during product development [WP-10-06] Integration and verification specification [WP-10-07] Integration and verification report
Clause 11 Cybersecurity validation	[WP-11-01] Validation report
Post-development phases	

Clause 12 Production	[WP-12-01] Production control plan
13.3 Cybersecurity incident response	[WP-13-01] Cybersecurity incident response plan
13.4 Updates	None

Sub-clauses	Work products
14.3 End of cybersecurity support	[WP-14-01] Procedures to communicate the end of cybersecurity support
14.4 Decommissioning	None
Threat analysis and risk assessment methods	
15.3 Asset identification	[WP-15-01] Damage scenarios [WP-15-02] Assets with cybersecurity properties
15.4 Threat scenario identification	[WP-15-03] Threat scenarios
15.5 Impact rating	[WP-15-04] Impact ratings with associated impact categories
15.6 Attack path analysis	[WP-15-05] Attack paths
15.7 Attack feasibility rating	[WP-15-06] Attack feasibility ratings
15.8 Risk value determination	[WP-15-07] Risk values
15.9 Risk treatment decision	[WP-15-08] Risk treatment decisions

Cyber Assurance Levels

ISO21434 has utilized the concept of Cybersecurity Assurance levels (CALs) classification scheme that can be used to provide "Assurance" and "Trust" on the level of Cybersecurity embedded into the Vehicle components and the Vehicle itself.

Assurance is the measurement of correctness and a judgement of a system's effectiveness of security functionality. Basically, it is the Degree of Confidence in the Cybersecurity implementation of any system.

A Trusted System is designed and implemented in such a way that hardware, firmware, Operating System, and software together effectively support the security policy.

As per Annex E of the standard,

"The CAL can be assigned based on the maximum impact, and the attack vector of the relevant threat scenarios"

The below table gives an example of how CAL can be used to enumerate the security profile of a specific component.

Table 4 Example of CAL determination

		Attack Vector			
		Physical	Local	Adjacent	Network
Impact	Negligible	---[1]	---[1]	---[1]	---[1]
	Moderate	CAL1	CAL1	CAL2	CAL3
	Major	CAL1	CAL2	CAL3	CAL4
	Sever	CAL2	CAL3	CAL4	CAL4

Demonstration and Evaluation of Supplier Capability

A critical component and far-reaching component of ISO 21434 is the focus on supplier capability from an Automotive Cybersecurity perspective. Clause 7.2 of the standard defined the objectives for distributed cybersecurity as:

"The objective of this clause is to define the interactions, dependencies, and responsibilities for distributed cybersecurity activities between customers and suppliers."

This clause details the requirements to showcase evidence for cybersecurity compliance by a supplier for any component of the vehicle deemed in scope for compliance to the ISO 21434 standard. This evidence would include artifacts of the organization's capability concerning cybersecurity best practices in all areas including concept and design, development, post-development, and decommissioning.

This clause has two major implications for automotive manufacturers. First, OEMs and Suppliers need to ensure that they have an adequate Cyber Supply Chain Risk Management Program and second, Automotive Organizations must have specific evidence of the organization's capabilities concerning cybersecurity.

NIST Cyber Security Framework (CSF)

The National Institute of Standards and Technology (NIST) Cyber Security Framework (CSF)[5] was established in 2013 because of the Improving Critical Infrastructure Cybersecurity Executive Order issued by United States President Barack Obama. The NIST CSF aims to have organizations build their Cybersecurity Organizations based on the voluntary development of a risk-based cybersecurity framework. The NIST CSF has the following principles:

- Goal of improving critical infrastructure cybersecurity

- Applying the principles and best practices of risk management

---

[5] https://www.nist.gov/cyberframework

- Improving the security and resilience of critical infrastructure

The Framework provides a common taxonomy and mechanism for Organizations to:

- Describe their current cybersecurity posture

- Describe their target state for cybersecurity

- Identify and prioritize opportunities for improvement within the context of a continuous and repeatable process

- Assess progress toward the target state

- Communicate among internal and external stakeholders about cybersecurity risk.

The NIST CSF is divided into five "Function" areas and each Function is divided into separate categories and sub-categories. A comprehensive Cybersecurity Program could adopt the NIST CSF and develop relevant Cybersecurity Policies & Procedures to ensure all applicable Functions and Categories are adequately protected in the organization.

Table 5 NIST CSF Functions & Categories

Function Identifier	Function	Category Identifier	Category
ID	Identify	ID.AM	Asset Management
		ID.BE	Business Environment
		ID.GV	Governance
		ID.RA	Risk Assessment
		ID.RM	Risk Management Strategy
		ID.SC	Supply Chain Risk Management
PR	Protect	PR.AC	Identity Management and Access Control
		PR.AT	Awareness and Training
		PR.DS	Data Security
		PR.IP	Information Protection Processes and Procedures
		PR.MA	Maintenance
		PR.PT	Protective Technology
DE	Detect	DE.AE	Anomalies and Events
		DE.CM	Security Continuous Monitoring
		DE.DP	Detection Processes
RS	Respond	RS.RP	Response Planning
		RS.CO	Communications
		RS.AN	Analysis
		RS.MI	Mitigation
		RS.IM	Improvements
RC	Recover	RC.RP	Recovery Planning
		RC.IM	Improvements
		RC.CO	Communications

ENISA Good Practices for security of Smart Cars

ENISA is the European Union Agency for Cybersecurity and was established in 2004 to achieve a high and common level of cybersecurity across Europe. It was further strengthened by EU Cybersecurity Act of June 27, 2019.

As per ENISA[6], *"the European Union Agency for Cybersecurity contributes to EU cyber policy, enhances the trustworthiness of ICT products, services, and processes with cybersecurity certification schemes, cooperates with Member States and EU bodies, and helps Europe prepare for the cyber challenges of tomorrow. Through knowledge sharing, capacity building and awareness raising, the Agency works together with its key stakeholders to strengthen trust in the connected economy, to boost resilience of the Union's infrastructure, and, ultimately, to keep Europe's society and citizens digitally secure."*

In November 2019, ENISA published the ENISA good practices[7] for security of Smart Cars. This is a comprehensive report and focuses on automotive cybersecurity in a holistic way to ensure cyber best practices are adopted in automotive manufacturing. Below is an excerpt from the executive summary of the report:

*"In this report ENISA defines smart cars as **systems providing connected, added-value features in order to enhance car users experience or improve car safety**. It encompasses use cases such as telematics, connected infotainment or intra-vehicular communication. The report excludes Car-to-car as well as autonomous vehicles as these technologies are not in use today. Practices discussed in this report concern not only passenger cars but also commercial vehicles (such as busses, coaches etc) and aim to map the current threats that passengers and drivers are exposed every day to. The goal is to secure smart cars today for safer autonomous cars tomorrow."*

---

[6] https://www.enisa.europa.eu/about-enisa
[7] https://www.enisa.europa.eu/publications/smart-cars

UN ECE WP.29 R155 & R156 Cybersecurity Regulations

The United Nations Economic Commission for Europe (UNECE) was set up in 1947 as one of the five regional commissions of United Nations Economic & Social Council (UN ECOSOC). As per the UN ECE website, its mission[8] is defined as

"UNECE's major aim is to promote pan-European economic integration. UNECE includes 56 member States in Europe, North America and Asia. However, all interested United Nations member States may participate in the work of UNECE. Over 70 international professional organizations and other non-governmental organizations take part in UNECE activities."

WP.29 is a working party of the Sustainable Transport Division of UNECE and its formal name is the World Forum for Harmonization of Vehicle Regulations. It is the responsibility of WP.29 to manage the multilateral Agreements signed in 1958, 1997 and 1998 concerning the technical prescriptions for the construction, and approval of wheeled vehicles as well as their Periodic Technical Inspection. Furthermore, WP.29 must work to further develop, improve & amend UN Regulations, UN Global Technical Regulations and UN Rules within the framework of these three Agreements.

As per the UNECE website, the main objective of WP.29 is

"Overall, the regulatory framework developed by the World Forum WP.29 allows the market introduction of innovative vehicle technologies, while continuously improving global vehicle safety. The framework enables decreasing environmental pollution and energy consumption, as well as the improvement of anti-theft capabilities."

On 24 June 2020, two new UN Regulations were adopted by UNECE's World Forum for Harmonization of Vehicle Regulations (WP.29) that ensure a focus on cybersecurity for Connected & Autonomous (CAVs). These regulations are

R155[9] - Uniform provisions concerning the approval of vehicles with regards to cyber security and cyber security management system

---

[8] https://www.unece.org/mission.html
[9] https://unece.org/transport/documents/2021/03/standards/un-regulation-no-155-cyber-security-and-cyber-security

R156[10] - Uniform provisions concerning the approval of vehicles with regards to software update and software updates management system

These two regulations require that measures be implemented across four (4) distinct domains:

- Managing vehicle cyber risks
- Securing vehicles by design to mitigate risks along the value chain
- Detecting and responding to security incidents across vehicle fleet
- Providing safe and secure software updates and ensuring vehicle safety is not compromised, introducing a legal basis for so-called "Over-the-Air" (O.T.A.) updates to on-board vehicle software.

The regulations will apply to passenger cars, vans, trucks, and buses and will enter into force in 2022. It is important to note that these regulations do not stipulate a specific cybersecurity methodology but requires that such measures be implemented in the Automotive Supply Chain that secure the vehicle. This means that organizations can implement ISO 21434 or any similar methodology to ensure compliance with these regulations.

Transport Canada's Vehicle Cyber Security Guidance

In June 2020, Transport Canada has issued Vehicle Cyber Security[11] guidance. This document was developed with the input from a broad array of expertise around automotive cybersecurity and defines principles of automotive cybersecurity. As per Transport Canada

The principles within the Cyber Guidance encourage organizations to:

- *identify how they will manage cyber security risks;*
- *protect the vehicle ecosystem with appropriate safeguards;*
- *detect, monitor, and respond to cyber security events; and,*
- *recover from cyber security events safely and quickly*

---

[10] https://unece.org/transport/documents/2021/03/standards/un-regulation-no-156-software-update-and-software-update
[11] http://publications.gc.ca/site/eng/9.884523/publication.html

The below table outlines the principles defined in this guidance document:

Table 6 Transport Canada Key Principles

Clause	Key Principle
1.	ORGANIZATIONS SHOULD IDENTIFY AND MANAGE CYBER SECURITY RISKS
1.1	Cyber Security Governance
1.2.	Risk Management Frameworks
1.3	Supply Chain Security
2.	ORGANIZATIONS SHOULD PROTECT THE VEHICLE ECOSYSTEM
2.1	Layered Cyber Defenses
2.2	Privacy Protection
2.3	Information Protection Procedures
2.4	Training and Awareness Programs
3.	ORGANIZATIONS SHOULD DETECT, MONITOR AND RESPOND TO CYBER SECURITY EVENTS
3.1	Event Detection, Monitoring and Analysis
3.2	Security Audits
3.3	Vulnerability Management Plan
3.4	Incident Management and Response
4.	ORGANIZATIONS SHOULD RECOVER FROM CYBER SECURITY EVENTS SAFELY AND QUICKLY
4.1	Incident Recovery
4.2	Partnership Building and Information Sharing
4.3	Cyber Security as a process of Continuous Improvement

Trusted Information Security Assessment Exchange (TISAX)

Trusted Information Security Assessment Exchange (TISAX) compliance is another method for Automotive sector ensure cybersecurity is enabled in their organizational ecosystem. As the name suggests, TISAX is an inter-company exchange mechanism having at its core the Information Security Assessment developed by the German Association of the Automobile Industry (VDA). Although TISAX is globally recognized, it is specifically required to do business with German vehicle manufacturers.

TISAX defines maturity levels[12] to define the status of the organization on its cybersecurity journey. These maturity levels are:

- **Level 0 (Incomplete)**
  There is no process, or the process does not work.
- **Level 1 (Performed)**
  There is a process, and the result suggests it works, but the process is not documented, and nobody knows for sure why the process works.
- **Level 2 (Managed)**
  There are processes that work and are documented, but there are many different processes for the same objective.
- **Level 3 (Established)**
  There is a process that works and has documentation that is up-to-date and maintained.
- **Level 4 (Predictable)**
  Same as for level 3, plus the process is measured.
- **Level 5 (Optimizing)**
  Same as for level 4, plus dedicated staff is responsible for continual improvements.

ISO 27000 Series

Although ISO 27000 Series is not, strictly speaking, a framework for automotive cybersecurity, it is an important cyber framework as its main aim is to manage cybersecurity in an enterprise environment. The standard ISO 27001: Information Security Management Standard was jointly published in 2005 by the International Organization for Standardization (ISO) and the International Electrotechnical Commission(IEC). It was then revised in 2013.

---

[12] ENX TISAX Participant Handbook

It is important to understand the key difference between the two main documents of the ISO 27001 Series, ISO 27001 and ISO 27002. ISO 27001[13] is the actual standard which holistically defines how to manage information security in an organization through an Information Security Management System (ISMS). To achieve this objective, the standard specifies requirements to establish, implement, maintain, and continually improve an information security management system (ISMS).

ISO 27002, on the other hand, is a supplementary standard that gives details of the Information security controls to establish, implement, maintain, and continually improve the ISMS. These controls are given as one-liners in Annex A of ISO 27001. However, ISO 27001 explains the objective of each control in detail and how it can be implemented in an organization.

There are many standards and guidance documents that are part of the ISO 27000 family. A sample list is given below:

- ISO/IEC 27000:2018 - an overview and introduction to the ISO27k standards
- ISO/IEC 27001:2013 - the certifiable standard defining requirements for an ISMS
- ISO/IEC 27002:2013 - the code of practice for information security controls
- ISO/IEC 27003:2017 - guidance on how to implement ISO/IEC 27001
- ISO/IEC 27004:2016 - information security management measurement
- ISO/IEC 27005:2018 - information [security] risk management
- ISO/IEC 27009:2020 - producing sector- or industry-specific ISO27k standards
- ISO/IEC 27010:2015 - information security management for inter-sector & inter-organisational communications
- ISO/IEC 27011:2016 - guideline for telecommunications organizations (= ITU-T X.1051)
- ISO/IEC 27013:2015 - guidance on the joint implementation of both ISO/IEC 27001 (ISMS) and ISO/IEC 20000-1 (IT service management or ITIL)
- ISO/IEC TR 27016:2014 - the economics of information security management

---

[13] https://www.iso.org/isoiec-27001-information-security.html

139

- ISO/IEC 27017:2015 - information security controls for cloud computing (= ITU-T X.1631)
- ISO/IEC 27018:2019 - concerns Personally Identifiable Information in public cloud
- ISO/IEC 27021:2017 - the competencies, skills & knowledge required by information security management professionals
- ISO/IEC 27031:2011 concerns ICT resilience and recovery for business continuity
- ISO/IEC 27033:2010 - IT network security
- ISO/IEC 27034:2011 - Application security
- ISO/IEC 27035:2016 - Information security incident management
- ISO/IEC 27036:2013-2016 - Supplier relationships
- ISO/IEC 27037:2012 - Identifying, gathering, and preserving digital evidence
- ISO/IEC 27038:2014 - Redaction of digital documents
- ISO/IEC 27039:2015 - Intrusion Detection and Prevention Systems (IDS/IPS)
- ISO/IEC 27040:2015 - Storage security
- ISO/IEC 27102:2019 – Cyber Insurance
- ISO 27799:2016 - health sector specific ISMS implementation guidance based on ISO/IEC 27002:2013

There are many guidelines of the ISO 27000 series that would be coming out in the future. However, an important one to mention in this regard is

- ISO/IEC TS 27570 - Privacy guidance for Smart Cities

Payment Card Industry Data Security Standard (PCI DSS)

As we move towards more connectivity & use in the Connected & Autonomous Vehicles (CAVs), many ECU Apps, such as infotainment, will incorporate the feature to accept credit card payments for their services directly from the vehicle itself. This would mean that ECU App developers must incorporate the requirements for the Payment Card Industry Data Security Standard (PCI DSS)[14] in their development and operations profile.

Payment Card Industry Data Security Standard (PCI DSS) has been published by the Payment Card Industry Security Standards Council (PCI SSC), an organization formed by Visa, MasterCard, Discover Financial Services, JCB International and American Express. The aim of PCI SCC and the PCI DSS is simple and truly relevant in today's ecommerce world i.e., to secure credit and debit card transactions against data theft and fraud. This is achieved by allowing only those organizations to process credit or debit card transactions that have achieved PCI Certification.

What you Learned in this Chapter

The aim of this chapter was to explain Cyber Governance Frameworks. Some of the topics covered include:

- The purpose of the Cyber Governance Framework for an Organization
- The critical elements of Cyber Governance Frameworks including Policies, Standards, Procedures, Baselines and Guidelines which enable the structure of the Cyber Governance framework.
- The various types of policies which ensure the successful implementation of any Cyber Governance Programs in an organization.
- The major Cybersecurity standards, including:
    - ISO 21434 Road Vehicles Cybersecurity Engineering Standard
    - NIST Cyber Security Framework (CSF)
    - ENISA Good Practices for security of Smart Cars
    - UNECE WP.29 Cybersecurity Regulations
    - Transport Canada Vehicle Cybersecurity Guidance
    - Trusted Information Security Assessment Exchange (TISAX)
    - General Data Protection Regulation (GDPR)
    - Payment Card Industry Data Security Standard (PCI DSS)
    - Sarbanes-Oxley (SOX) 404
    - ISO 27000 Series (Information Security Management System)

---

[14] https://www.pcisecuritystandards.org/

Chapter 9: Cyber Supply Chain Risk Management

Chapter Overview

This chapter deals with Cyber Supply Chain Risk Management (C-SCRM) which is a critical component in ensuring the smooth functioning of the Automotive Supply Chain. A modern vehicle has more than 30,000 suppliers and the OEM is dependent on the smooth operations of each one of its suppliers to ensure manufacturing of quality vehicles. Therefore, it is imperative to have a comprehensive Cyber Supply Chain Risk Management (C-SCRM) Program that develops a holistic relationship looking at all aspects of Cybersecurity between the OEMs and the Suppliers. In the CAV ecosystem, it becomes even more important to effectively manage Supplier Cyber Risk as many of the Electrical / Electronics components would be dependent on Over-the-Air (OTA) updates from the component maker. Having such a Supplier Cybersecurity Maturity Model shall ensure that Cyber incidents are limited and there is proper communication between the OEM and its suppliers to effectively manage any Cyber incidents.

Cyber Supply Chain Risk Management (C-SCRM) Program

The challenge for any organization is to effectively manage the Cyber risk from its suppliers. This is especially challenging for an Automotive Manufacturer due to the unique nature of the automotive industry. The automotive supply chain is long and complex with a large number and diverse nature of suppliers spread over many different jurisdictions. With the advent of the Connected and Autonomous Vehicle (CAV), this supply chain has become even more complex as it now includes technology vendors as well as Cloud Apps. Furthermore, OEMs and other Automotive entities are reliant on third parties such as data centers and technology service providers. A break in the chain at any point due to a cyber incident, whether an OEM, a Supplier, or a Technology Service Provider, can be disastrous for the end-consumer i.e., the passenger in the vehicles.

To meet this challenge, a comprehensive Cyber Supply Chain Risk Management (C-SCRM) Program needs to be established at any automotive organization. The major goal of this program would be to mitigate the potential for a large and/or public cybersecurity breach of an OEM that could be attributed to the deficiencies in the cyber practice of a vendor. Other benefits incurring from the Supplier Cyber Risk Management Program would include:

- A comprehensive collaborative Cyber Strategy for the entire OEM eco-system.
- A commercial relationship between the Automotive Manufacturer and its suppliers that enhances the value-add to all parties involved.
- A decrease in risk exposure to the supplier due to service failure or non-compliance.
- A reduction in the supplier not being able to adhere to contractual obligations.
- An increase in supplier competitiveness through better cyber hygiene.
- Effective & robust Cyber Incidence Response through-out the supply chain.

An important challenge while developing an effective Cyber Supply Chain Risk Management (C-SCRM) Program is the fact that it requires Cyber SMEs with a unique skillset. These Cyber SMEs need to understand both the Supplier eco-system and the possible threats to that eco-system. In addition, these Cyber SMEs would be carrying out continuous cyber assessment of its suppliers and need to ensure effective communication of perceived threats & the controls that need to be put in place to mitigate against these threats. This means that Threat Modeling for the whole eco-system would become an important component of such a holistic program.

Supplier Selection

Cyber risk should be considered as an essential component of the Supplier Selection criteria for an Automotive manufacturer. This is in addition to the core set of supplier risk criteria that typically includes the following:

- Financial risks

- Geolocation risk
- Business continuity and time to recovery risks
- Operational risks (Time to Delivery, Quality, Performance, Cost & Capacity)
- Safety risks

These risks criteria considered during the Supplier Selection process are a good indicator of whether a supplier or sub-supplier can deliver components of the vehicle on time and as expected. Standards like ISO 21434 are in fact providing guidelines to evaluate risk beyond the supplier level i.e., define Cyber Assurance Levels (CALs) for the individual components. This means that these companies are not only mapping individual suppliers but must also map & track Cyber risk in individual components, products, and equipment.

To assess the cyber risk of the supply chain, automotive cybersecurity SMEs at Automotive Manufacturers need to ask some of the below questions:

- Has Cyber been defined as an area of concern at the supplier?
- What definition of Cyber is being utilized? Is Cyber only looked at from an IT perspective or is the definition more holistic including Vehicle, V2X, OT, IT, Edge, Cloud & Physical security perspectives?
- What are the HR policies at the supplier with regards to their employees, especially about individuals who are at critical positions with access to the data, systems, or facilities of their customers? How is the organization protecting against insider attacks?
- How well do the suppliers themselves scrutinize their service providers?
- How well do the suppliers themselves assess their own components, products, and software for cyber risk? Any Electronic component to be supplied for CAVs would be of particular concern.
- Is the Supplier compliant to any generic Cybersecurity standard or framework such as ISO 27001 or NIST CSF?
- Is the Supplier compliant to any Automotive Cybersecurity standard or framework such as ISO 21434 or TISAX?

Supplier Cyber Requirements

A comprehensive Cyber Supply Chain Risk Management (C-SCRM) Program should define Supplier Cyber Requirements that can be provided to any new or existing suppliers. These requirements should also be used as a guideline to assess the supplier on a periodic basis through the Cyber Supply Chain Risk Management (C-SCRM) Program.

Below are some of the possible Cyber requirements for a Cyber Supply Chain Risk Management (C-SCRM) Program at an Automotive Organization

- Definition of Cyber
- Cyber Governance
    - Cyber Strategy
    - Ownership of Cyber Mandate
    - Cyber standards or frameworks implemented
- Risk Assessment Methodology
- Manufacturing Cybersecurity
    - Operational Technology
- In-Vehicle Cybersecurity
    - Architecture
    - Encryption
    - Authentication
    - Software Bill of Material
- V2X Cybersecurity
    - Encryption
- Security by Design
    - Secure Software Development
    - Secure Architecture
- Asset Management
- Incident Management
    - Continuous Monitoring
    - Vulnerability Management
    - Over-The-Air Updates
- Information Protection
    - Data Classification
    - Privacy & Tracking
- Physical and Environmental Security

- Personnel Security
- Supply Chain security
    - Contractual terms on Cybersecurity
    - Supplier Cyber Interface Agreements
    - SaaS apps (Cloud) Security

Supplier Cyber Interface Agreements

Clause 7 of ISO 21434 specifically mandates Supplier Cyber Interface Agreements[1] between the OEM and its suppliers. The aim of these agreements is two-fold: First, to ensure an understanding of Cyber responsibilities of both the Supplier and the Manufacturer during the development of a connected component and two, to develop an effective communication channel to handle cyber incidents during the operations of the connected component. The Supplier Cyber Interface Agreements is key to managing the relationship with suppliers, both internal and external, and should be part of a holistic Supplier Cyber Risk Management Program.

The Supplier Cyber Interface Agreement should look at the below areas:

- Evidence of Organization's capability concerning cybersecurity
- Evidence of continuous cybersecurity activities
- Summary of previous cybersecurity assessments
- Results of previous cybersecurity audits
- Evidence of an operational Cyber Security Management System (CSMS)
- Evidence of components of a Management System such as change management & documentation management

The Supplier Cyber Interface Agreement should also define the responsibilities for the development and operations of any component between the OEM and its suppliers.

Supplier Cyber Assessments

An organization must decide the methodology through which it would carry out Supplier Cyber Assessments. It would have to develop a Supplier Threat Analysis

---

[1] ISO 21434 Road Vehicles Cybersecurity Engineering Standard

& Risk Assessment (TARA) to ensure that each supplier's Cybersecurity capabilities are clearly recorded and understood.

A Supplier Cyber Assessment should also focus on the Cyber Standards & Frameworks which its suppliers have already implemented in their environment or should adhere to as per the OEMs Cyber requirements. As mentioned in chapter 8, there are many Cyber Standards & Frameworks which might be applicable to an Automotive Organization:

- ISO 21434 Road Vehicles Cybersecurity Engineering Standard
- NIST Cyber Security Framework (CSF)
- ENISA Good Practices for security of Smart Cars
- UNECE WP.29 R155 & R156 Cybersecurity Regulations
- Transport Canada Vehicle Cybersecurity Guidance
- Trusted Information Security Assessment Exchange (TISAX)
- General Data Protection Regulation (GDPR)
- Payment Card Industry Data Security Standard (PCI DSS)
- Sarbanes-Oxley (SOX) 404
- ISO 27000 Series (Information Security Management System)

Many of these frameworks have similar or overlapping requirements which can be met through the implementation of the same controls. Therefore, it might be useful to carry out a control mapping exercise between the various frameworks. This will help validate the requirements that have already been met for specific frameworks while implementing any new frameworks.

Cloud Apps Risk Assessment

Developing a Cloud Apps Risk Assessment Policy is an essential component of today's Cyber Supply Chain Risk Management (C-SCRM) Program. Having such a policy in place ensures that the organization can regulate the use of third-party applications and infrastructure resources that users access in the Cloud. Cloud suppliers can be divided into major categories: Cloud platforms such as AWS, Azure & Google Cloud and the suppliers of thousands of SaaS apps which are being utilized in any enterprise. It is critical to differentiate between both these types of Cloud Apps as the cybersecurity approach to protect each set of these apps is completely distinct.

Cloud Platforms are used to host internal applications owned by the organization itself in the Cloud. Companies like Amazon, Microsoft and Google have made considerable investments in their Cloud Hosting platforms. In addition, they provide tools to their customers, the enterprises themselves, to secure their data. Thus, an enterprise has considerable control on the security profile of their hosted applications & data on these platforms.

The second category i.e., the SaaS apps, are much less secure and much more difficult to protect. This is because the enterprise has absolutely no control on the cybersecurity of their data on these apps and must rely on tools such as CASBs (Cloud Access Security Brokers) to protect their data going to these apps. An important point to note here is that these SaaS apps can also be further divided into two sub-categories. First, Sanctioned apps where the enterprise has a contractual relationship with the vendor of that app. Examples of these apps are Microsoft O365, ServiceNow and SalesForce. The second sub-category are the so-called Tolerated Apps. This is the vast majority of Shadow-IT which has become prevalent in any organization over the last few years. These apps need to be identified and data sent to them monitored & protected. Finally, there would be Unsanctioned Apps that must be blocked from being used in the organization.

The main objective of Cloud Apps Risk Assessment Policy is to establish a standard of practice for the procurement, risk evaluation, and use of Cloud Apps that your organization relies on each day. By applying this policy, the organization can establish a level of Cyber guidance when procuring Cloud, especially, SaaS Apps, managing users, protecting data, and securing assets in the Cloud. Such a policy will also ensure that the organization's confidential information and data which is held by third parties is protected and remains uncompromised.

The same would apply in the CAV eco-system for any Cloud Apps being used by the CAV. For all Cloud Apps being accessed from the CAV, Sanctioned, Tolerated and Unsanctioned Apps must be identified and data being sent to these protect to ensure that PII is shared and stored only at authorized resources.

Supplier Cybersecurity Maturity Model (SCMM)

As mentioned earlier in this chapter, the automotive supply chain is complex and has thousands of suppliers. Also, as the Automotive Manufacturing sector is only now starting out on this journey, OEMs and the Tier 1 Suppliers must ensure that their partners in the supply chain are educated and enabled to move forward together on this journey. Only a comprehensive and long-term Supplier Cyber Risk Management Program can ensure that data protection is available throughout an integrated supply chain.

To achieve the above-mentioned objective, a good model is to develop a Supplier Cybersecurity Maturity Model (SCMM). German automakers have already implemented this model through the Trusted Information Security Assessment Exchange (TISAX) which defines maturity levels of an organization with respect to their Cyber capabilities. OEMs and Suppliers can develop and follow a similar internal rating through a Supplier Cybersecurity Maturity Model (SCMM) that helps to guide suppliers to enhance their cybersecurity capabilities.

Supplier Cyber Management Best Practices

Automotive organizations should have clearly defined Supplier Cyber Management Best Practices that help the internal stakeholders as well as the suppliers realize the criticality of Cyber in the organization's supply chain. Some of the cyber practices that have helped businesses manage their suppliers more effectively include:

- Cyber is about brand integrity and reputation risk of the OEM which is inherently tied to those of its suppliers
- The integrated and connected nature of the supply chain, as well as connected components in the vehicle demand life-cycle threat modeling. This is a must to proactively detect vulnerabilities and mitigate potential threats in the supply chain.
- Standard Cybersecurity Terms and Conditions must be enumerated and included in all requests for proposals (RFPs) and contracts.
- All sourcing and procurement decisions should ensure inclusion of all stakeholders including Infrastructure cybersecurity, Product cybersecurity, Operations cybersecurity, and any other relevant departments.

- Any exceptions to cybersecurity provided to a supplier should be through a formal process with approval from C-level officers who are fiduciary responsible for the business impact.
- All suppliers should be continuously audited to ensure compliance with the OEM's Cyber Supply Chain Risk Management Program.
- Companies using on-site verification & validation for supplier cyber self-assessments must ensure that the personnel auditing its suppliers understand the eco-system of the supplier.
- Mentoring and training programs focused on enabling cybersecurity must be offered to suppliers.
- Approved vendor lists should only include suppliers that have undergone the initial onboarding process for the Cyber Supply Chain Risk Management Program.
- Quarterly reviews and annual meetings of business & cyber decision-makers of both the manufacturer and the supplier must take place to highlight and understand the common business needs and cyber concerns.
- Supplier representation is a must for managing & mitigating Cybersecurity Incidents.

Supply Chain & Cyber Incident Management

A critical area to consider is the involvement of suppliers in the management of a Cyber incident. It is imperative that supplier cybersecurity representatives be part of the OEM Cyber Incident Response Team (CIRT) and have active involvement before, during & after a cyber incident. Incident response planning should include active involvement from the cybersecurity representatives of all suppliers. Communications during any cyber breach should ensure that suppliers are fully informed of the details of the specific incident and the steps being taken to mitigate the incident. The whole supply chain should benefit from the lessons learnt during the cyber incident. Any updates in the organizations Policies & Procedures because of the cyber incident should also be communicated to all the suppliers.

Only by having such a dynamic, close, and robust Cyber Incident Management relationship with its suppliers can an Automotive organization ensure that it can react quickly and decisively to any cyber incidents happening on its vehicles and infrastructure.   However, if this approach is lacking, an organization shall face severe reputation & financial loss in case of a major cyber breach or DoS attack.

What you Learned in this Chapter

This chapter highlighted the following:

- Importance of Supplier Cyber Risk Management in today's automotive supply chain
- Development of a comprehensive Cyber Supply Chain Risk Management (C-SCRM) Program
- Enabling a Supplier Cybersecurity Maturity Model to help create a path for Suppliers to gradually improve & enhance their Cybersecurity
- Defining Supplier Cybersecurity best practices
- Ensuring proper communication between the OEM and its suppliers to effectively manage any Cyber incidents

Chapter 10: Vehicle Operations Cyber Risk Management

Chapter Overview

This chapter focuses on various techniques that must be employed to ensure successful cyber risk mitigation for the operational environment. As previously described in this book, cyber initiatives across several domains are required to secure an automotive organization. These include establishing a Cyber Governance program for the organization, understanding the threats & risks to the organization, enabling adherence to specific Cyber frameworks, and developing a Cyber Supply Chain Risk Management Program (C-SCRM). However, these are just the building blocks of the entire CSMS and once the foundation is built, it must be ensured that the organization's data assets are protected against cyber threats in the operational environment. This can only be achieved through robust and continuous cyber risk mitigation strategies. These strategies include Security by Design, Defense in Depth, Vulnerability Management Program, Vehicle Cyber Monitoring and Vehicle Cyber Incident Response Management.

Operations Cyber Risk Mitigation

With the advent of Connected & Autonomous Vehicles (CAVs), the existential dimensions of Cyber Risk have become manifest to OEMs, Tier1s and Tier2s. This risk is not only inherent in the conceptual and development phases of the vehicles but also needs to be addressed during the operational phase of the Connected Vehicle. Upcoming Cyber regulations and frameworks such as UNECE WP.29 as well as ISO 21434 also highlight the criticality of mitigating cyber risk during the operations of the vehicle. All this entails that the Chief Information Security Officer (CISO) must develop a holistic Cyber Risk Mitigation approach for the entire operations of the vehicle.

The above also implies that the Chief Information Security Officer (CISO) must collaborate with a diverse group of stakeholders to formulate this holistic Cyber

Risk Mitigation approach. This diverse group of stakeholders includes various internal organizational elements i.e., Business, Manufacturing, Legal, IT & Product Development. However, the CISO must also ensure that the Supplier perspective is similarly part of this holistic approach to protecting the organizational mobility assets from Cyber Risk as a cyber incident in a vehicle might happen due to vulnerabilities in the supply chain. Cyber Risk Mitigation necessitates a true partnership between all these stakeholders as no single entity has the complete perspective and knowledgebase to be effective in the Cyber domain, especially in the Automotive sector.

Based on the complex technological landscape of automotive cybersecurity, numerous actions are needed to address operational risk associated with technology in this sector. This chapter has encapsulated all these actions into five main areas of focus for the Vehicle Cybersecurity SME. These include:

- Security by Design
- Defense in Depth
- Vulnerability Management Program
- Vehicle Cyber Monitoring
- Vehicle Cyber Incident Response Management

Security by Design

As previously mentioned in Chapter 3, Security by Design is an essential tool in the Cybersecurity SMEs toolkit. Sadly, this is too often not followed and often only after a major malicious incident is the Cybersecurity SME brought in to secure the crown jewels. Even then, most organizations do not start building Cybersecurity at the earliest phases of the software development effort and most only react to the Cybersecurity SMEs lead. Only a major culture shift can ensure that Cyber is top of mind and Security by Design is followed.

In the automotive sector, having a Security by Design approach is integral to the smooth functioning of the modern vehicle in the operations phase. This is because there are between 20,000 to 30,000 suppliers in any modern vehicle. In addition, a Connected Vehicle would have between fifty to a hundred ECUs. A Security by Design proactive approach can ensure prevention of vulnerabilities in the connected vehicle components that might be used by attackers to gain access into the entire vehicle. ISO 21434, the Road Vehicles Cybersecurity Engineering standard, specifically outlines the cybersecurity requirements for the "Concept Phase" and the "Development Phase" to ensure various cybersecurity requirements are met.

Concept Phase

Clause 9.2 of the ISO 21434 standard defines the following cybersecurity objectives for the Concept Phase[1] of any Vehicle item or component:

a) define the item, its operational environment and their interactions in the context of cybersecurity;

b) specify cybersecurity goals and cybersecurity claims; and

c) specify the cybersecurity concept to achieve cybersecurity goals

Thus, ISO 21434 clearly outlines that cybersecurity goals should be defined at the very beginning of the development of any component to be incorporated in the Connected Vehicle. Inherent in point b above is the fact that a Threat Analysis & Risk Assessment (TARA) needs to be undertaken even at this stage to comprehend the threats and the risks to the component and its operational environment. Any relevant cybersecurity requirements must also be enumerated and documented at this stage to ensure that cyber controls are implemented in later stages of the product development.

Development Phase

Clause 10.2 of the ISO 21434 standard defines the following cybersecurity objectives for the Development Phase[2] of any Vehicle item or component:

a) define cybersecurity specifications;
b) verify that the defined cybersecurity specifications conform to the cybersecurity specifications from higher levels of architectural abstraction;
c) identify weaknesses in the component; and

---

[1] ISO 21434 Road Vehicles Cybersecurity Engineering Standard
[2] ISO 21434 Road Vehicles Cybersecurity Engineering Standard

d) provide evidence that the results of the implementation and integration of components conform to the cybersecurity specifications.

As is obvious from the above-mentioned objectives of this phase, only a culture change in the automotive manufacturing environment can guarantee that the above goals are being pursued during the Development phase of the item or component. This means that the software development team for any automotive component must understand the principles of secure coding, undergo regular trainings in this area, carry on cyber testing during the QA phase and communicate effectively with the wider cyber stakeholders of the vehicle item or component. This would be even more challenging when the Vehicle item or component such as a software module, is being developed by a Tier 2 supplier to be integrated into a Tier 1 ECU to be used by an OEM in an upcoming new model of a vehicle.

Defense-in-Depth Architecture

As mentioned in Chapter 3, "Defense in Depth" is the concept of having multiple layers of preventive & detective mechanisms implemented to protect valuable data and information. The main objective of "Defense in Depth" is to ensure that a hacker would have to spend time & effort to get to the stage where the target organization would be seriously impacted because of this specific security breach. In addition, such a layered defense would give the enterprise a good chance to prevent the attack all together because If one layer were overcome, there would be another that would thwart the attack. This would also allow the breach to be detected and allow for necessary incident response actions to be carried out.

This Defense-in-Depth approach with intentional redundancies must be architected in the design stage of the project. This is necessary to addresses the many different attack vectors and increase the security of the complete system. For a Connected Vehicle, this Defense-in-Depth Architecture needs to be developed for the entire automotive eco-system considering the attack vectors in many different zones.

These zones are:

- Vehicle Zone
- Enterprise Zone
- Production Zone
- Cloud Zone
- Supplier Zone
- SmartCity Zone

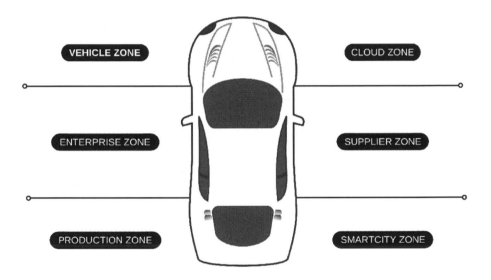

Figure 22 Cybersecurity definition for CAVs

Vehicle Zone

The Vehicle Zone will focus on cybersecurity controls that cover the design, development, operations & decommissioning of the Connected Vehicle itself. Cybersecurity controls in the vehicle include:

- Cyber Threat Analysis & Risk Assessment of the Vehicle itself
- In-Vehicle Secure Architecture elements such as segmentation & firewalls
- Secure Software development of all the ECUs in the vehicle
- Authentication & Identity Management

- In-Vehicle Encryption
- Cyber monitoring of traffic to and from the vehicle
- Vulnerability Management & Patch Updates
- Integration & data flow inside the vehicle
- Vehicle Incident Response
- Third-party In-Vehicle App Security

Enterprise Zone

The Enterprise Zone would cover the on-prem elements of the Automotive organization's Information Technology (IT) infrastructure. These includes the cell phones, workstations, servers, routers, firewalls, network cabling, IP Telephony, Call centers and Data centers that are involved in enabling data sharing & communication in a modern enterprise. Many frameworks are available such as NIST CSF & ISO 27001/2 that can be used to build controls in this zone.

Production Zone

The Production Zone is focused on enabling Cyber controls in the manufacturing of the vehicle and its components. As described in Chapter 5, legacy OT (Operational Technology) Systems have not been designed or implemented with security in mind. They are often proprietary and untested and do not work well with newer versions of Operating Systems. They do have strengths such as having a high degree of Availability and a high degree of Authorization. However, there are many significant weaknesses in OT Systems from the Cyber perspective. The two primary ones being allowing Total Authorization and Total Trust. In addition, their integration with IT and remote working environments have further weakened the defenses of the Production Zone.

Thus, there is a requirement to carry out a thorough Threat Analysis & Risk Assessment for the Production Zone and comprehensive Cyber controls need to be implemented in this Zone to ensure that critical assets, especially data, is protected from cyber attacks & incidents.

Cloud Zone

There are two important Cyber elements of focus in the Cloud Zone. The first are the applications which are owned by the OEM and are hosted in the major Cloud platforms such as AWS, Azure & GCP. These are, in effect, apps whose cybersecurity profile can be controlled and managed through the security tools provided by the Cloud platforms themselves. Thus, from a cyber perspective, the risks to these apps can be ascertained and relevant controls implemented.

However, the bigger challenge would be the other Cyber area of focus, i.e., securing the third-party apps in the Cloud which the OEM does not own or control in any way. Many organizations do not even know the apps that are being used in their traditional IT eco-system (Shadow IT). This inconspicuousness would become even more problematic when these apps are being installed by commuters directly into the vehicle interface itself. To visualize this challenge, the vehicle needs to be re-imagined as we use the cellphone currently. Any downloaded app with a vulnerability could be a gateway into the connected components itself and ultimately, provide hackers with a way to harvest PII and to disrupt transportation. With millions of Connected Vehicles and billons of apps being downloaded and used in these Connected Vehicles, the scale and breadth of this challenge becomes obvious.

Supplier Zone

An Automotive Organization must effectively manage the Cyber risk from its suppliers. Obviously, this is extremely complicated as the automotive supply chain is long and complex with a large number and diverse nature of suppliers. As detailed in Chapter 9, the solution here is to establish a comprehensive Cyber Supply Chain Risk Management Program. The major goal of such a program would be to mitigate the potential for a large and/or public cyber security breach that could be attributed to the cyber practice failings of a vendor. It would also enable a robust Cyber Incidence Response when a cyber attack takes place.

A comprehensive and effective Cyber Supply Chain Risk Management Program has many benefits including:

- A comprehensive collaborative Cyber Strategy for the entire OEM eco-system

- A more positive commercial relationship between the Automotive Manufacturer and its suppliers
- A decrease in risk exposure to the supplier due to service failure or non-compliance
- A reduction in the supplier not being able to adhere to contractual obligations
- An increase in supplier competitiveness through better cyber hygiene
- Effective & robust Cyber Incidence Response through-out the supply chain

SmartCity Zone

The SmartCity Zone would concentrate on the V2X component of the Connected Vehicle. The Cyber challenges of enabling the Vehicle's communication to SmartCity elements such as Intelligent Transport Systems (ITS) need to be understood and developed into a comprehensive control framework. Areas such as identity management, encryption, cyber monitoring, and data leakage are elements that need to be considered to ensure data protection & privacy. Currently, there is no single cyber framework established to ensure that Cyber Governance is incorporated in the policy discussions of SmartCities. This aspect of SmartCities needs to be integrated in the overall structure of SmartCities to ensure safety & security of individuals living in this Connected World of the Future.

Vehicle Vulnerability Management Program

With fifty to hundred ECUs per vehicle and millions of lines of code, establishing a well-defined Vehicle Vulnerability Management Program is a must for any OEM. In the days of Over-the-Air (OTA) updates, a Connected Vehicle can be hacked effortlessly if there are exploitable vulnerabilities present in the code. Thus, this concern has been specifically highlighted in the UN ECE WP.29 Cybersecurity regulations[3] as a distinct discipline to be met by the Automotive Manufacturing community. The regulation specifically states:

---

[3] https://www.unece.org/info/media/presscurrent-press-h/transport/2020/un-regulations-on-cybersecurity-and-software-updates-to-pave-the-way-for-mass-roll-out-of-connected-vehicles/doc.html

"Providing safe and secure software updates and ensuring vehicle safety is not compromised, introducing a legal basis for so-called "Over-the-Air" (O.T.A.) updates to on-board vehicle software."

To build a comprehensive Vehicle Vulnerability Management Program, an Automotive organization must have a process to assess, mitigate and report on any security vulnerabilities that exist in any software code in a vehicle's ecosystem. Any such Vehicle Vulnerability Management Program must also have an effective process to manage responsibility and communication for the various elements of the vulnerability management lifecycle. This is especially critical because of the challenges in vulnerability management dependent on the supply chain. An example of this is a software that might have been developed by a Tier 2 for an Infotainment system built by a Tier 1 incorporated in a vehicle by an OEM. This means that the OEM would be responsible for ensuring that the vulnerabilities in a software developed by a Tier2 does not have a vulnerability that can lead to hacking of a vehicle manufactured by that OEM and must work with the supply chain to discover, identify and patch these vulnerabilities.

Any Vulnerability Management Program must have the below distinct elements:

- Software Build of Material (SBOM)
- Vulnerability Scanning
- Penetration Testing
- Patch Management

Software Build of Material (SBOM)

To effectively operate the Vehicle of the Future, it is critical to build a Software Bill of Material (SBOM). As vehicles become more digitized, there would be hundreds of apps with millions of lines of code installed in them. This software would be from many different suppliers and Over-The-Air (OTA) patch updates would have to be carried out on a period basis to secure these apps. Thus, it is critical to keep track of the software installed on the vehicle. It is for this very reason that UNR 156 has deemed an SBOM an essential requirement.

An SBOM is a formal, machine-readable inventory of software components and dependencies, information about those components, and their hierarchical relationships. Such a comprehensive inventory is an essential aspect of safety for the technology supply chain of any vehicle and would represent the embedded software, firmware, and microcode inside the vehicle. Having such an SBOM would ensure that vehicle manufacturer have an UpToDate snapshot of all software code present in the vehicle itself. Having such an UpToDate snapshot will enable risk mitigation for the vehicle by identifying any existing code that needs to have updated patches.

There is significant work being by Auto-ISAC[4] in this area to ensure that SBOM best practices for the automotive sector are being clearly highlighted to the automotive sector. However, OEMs and suppliers must take the lead on this activity and ensure that their automotive software supply chain is clearly identified, and due care & due diligence carried out as per C-SCRM.

Vulnerability Scanning

Vulnerability Scanning is the process by which software is scanned for vulnerabilities and reports generated to mitigate them. A vulnerability scanner is a tool used to identify and create an inventory of all assets in scope. In the traditional IT world, these assets included systems such as workstations, desktops, virtual machines, containers, firewalls, switches, and printers. The vulnerability scanner would then go through the effort of discovering the operating system on the target asset, open ports of communication and any user accounts that might have been created to access that target system.

The final output of a vulnerability scanner also includes any vulnerabilities found in the identified assets. These are found by validating each item in the inventory of the vulnerability scanner to the vulnerabilities found in a pre-defined database built into the vulnerability scanner. The identified vulnerabilities are highlighted as high, medium, or low based on the severity of the vulnerability found. These can then be mitigated by applying patches to the system.

The challenge for the automotive manufacturing world is that, currently there are no industry-recognized vulnerability scanners made specifically for vehicle ECU scanning. Although traditional vulnerability scanners from the IT world can be used to a certain extent, these would not have the pre-defined databases to

---

[4] https://automotiveisac.com/

identify automotive OS such as Blackberry QNX or the vulnerabilities found in thousands of ECUs being developed for various components of the vehicle.

Penetration Testing

Penetration Testing is more intrusive than Vulnerability Scanning. In Penetration Testing, the vulnerabilities found through the process of vulnerability scanning are used to exploit and hack into the target systems. This means that Penetration Testing can be very disruptive to production systems and can even bring down an active system. Therefore, Penetration Testing must be done with great care on a vehicle and under very controlled parameters.

Below are the various phases of Penetration Testing:

- Information Gathering
- Reconnaissance
- Discovery & Scanning
- Vulnerability Assessment
- Exploitation
- Vehicle PenTest Report
- Report Utilization

In the domain of Automotive cybersecurity, the "Information Gathering" phase would be defined by gathering information about the in-scope target of the Penetration Test. This would include the type of vehicle such as SUV, Saloon, Truck or Bus; the make of the vehicle such as OEM, brand, and model; and any owner information available. "Reconnaissance" would include any other relevant information that can be gathered such as the area or route where the vehicle normally operates. "Discovery & Scanning" would use a vulnerability scanner to get initial information such as target OS, ECUs exposed or open ports. "Vulnerability Assessment" phase would actually identify the potential vulnerabilities which are available in the exposed ECUs to hack into the Vehicle while the "Exploitation" phase is the actual step where the hacker will gain access into the Vehicle itself. Once the Penetration Test is completed, "Vehicle PenTest Report" phase shall include analyzing the information generated from the Penetration Test and documenting it as a formal report. Finally, this report is used in the "Report Utilization" phase to enhance the cybersecurity controls of the vehicle.

Patch Management

A comprehensive and effective Patch Management process must be established by an Automotive Manufacturer to work together with its supplier. The goal of this Patch Management process must be to ensure that any code changes are installed on any specific ECU only through a structured process. This means that after the code for a specific ECU is acquired from the development team/supplier, it is first tested in staging environments and only after validating its impact, it should be applied on the ECUs in Production. This Patch Management process should also be able to handle patches from multiple suppliers concurrently while determining the sequence and priority of the patches. With millions of lines of code present in the modern vehicle, enabling ECUs to stay updated on existing & new patches is a complex undertaking and needs to be effectively managed.

Another area of concern for Patch Management in the modern vehicle is the mechanism through which the patches are delivered to the vehicle. Many automotive organizations still manage patch updates through a process where the owner takes the vehicle to the dealer for providing updates through the OBDII port. However, there is a growing trend to provide patches through Over-the-Air Updates. By its very nature, OTA Updates can be hijacked through a Man-in-the-Middle (MITM) attack. Thus, OTA updates must be secured through effective authentication & encryption methodology and monitored for cyber events.

Company Highlight: CyTaara Inc.

CyTaara secures the Technology Supply Chain

Core Mission:

Ensuring 'Security by Design' by leveraging the latest technologies to protect the safety and security of our customers and their critical infrastructure. This is done by integrating security best practices into all stages of the development lifecycle from training to development to maintenance

Vision:

To transform traditional software development into an innovative and adaptive secure software development life cycle for all our users by incorporating 'Security by Design'

www.cytaara.com

Vehicle Security Operations Center (VSOC)

As mentioned at the start of this book, Vehiqilla Inc.[5] was specifically founded to enable cyber monitoring to the Connected Vehicle. During the development of the Vehiqilla Cyber FleetSOC Center, all challenges related to establishing an effective Vehicle Security Operations Center (VSOC) were explored, understood, and resolved. This meant an in-depth understanding of all aspects of Vehicle Cyber Monitoring was developed by the entire team which is leveraged here.

The criticality of Vehicle Cyber Monitoring is highlighted by its emphasis on the new regulations & best practices being developed for automotive cybersecurity. Starting with the UNECE WP.29 Cybersecurity regulations, Cybersecurity Monitoring is one of the four distinct disciplines for which cybersecurity measures must be implemented by the automotive ecosystem. Specifically, it is said that cybersecurity measures[6] must be implemented for

- *"Detecting and responding to security incidents across vehicle fleet"*

Also, Best Practice 3.1 of Transport Canada's Vehicle Cybersecurity Guidance[7] is defined as

- *"Event Detection, Monitoring & Analysis"*

Finally, Clause 7.3 of the ISO 21434[8] also emphasizes Cyber Monitoring and mentions

- *"Cybersecurity monitoring (see 7.3) collects cybersecurity information on potential threats, vulnerabilities, and possible 598 mitigations for items and components to avoid known issues and to address new threats and can serve as the input for 599 vulnerability management and cybersecurity incident response activities."*

---

[5] https://vehiqilla.com/
[6] https://unece.org/press/un-regulations-cybersecurity-and-software-updates-pave-way-mass-roll-out-connected-vehicles
[7] https://tc.canada.ca/en/road-transportation/innovative-technologies/automated-connected-vehicles/connected-automated-vehicle-safety-what-you-need-know#Tools-and-resources
[8] ISO 21434 Road Vehicles Cybersecurity Engineering Standard

This highlights that Vehicle Security Operations Centers (VSOCs) must be created to "detect, monitor & analyze" cyber events to the Connected Vehicle during the operations phase of that vehicle. A VSOC must

- Get relevant logs as input from vehicles being monitored
- Have a database of Vehicles and ECUs
- Have details of vulnerabilities found in various ECUs
- Have details of breaches that affected various ECUs
- Have holistic cyber risk defined for all ECUs & Suppliers
- Have effective communication methods with Vehicle/Fleet Owners
- Have capability to effectively analyze data coming in through AI-based tools
- Have SMEs who understand Automotive Cybersecurity

Figure 23 The need for a VSOC

Vehicle Cyber Incident Response Management

Incidence Response Planning is a critical component of any Risk Management Plan. The main goal of Cyber Incidence Response Planning is to first detect & identify a cyber incident and then, based on the type & severity of the incident, effectively respond and recover from that cybersecurity incident. Having a robust and well-tested Cyber Incident Response Plan would limit significant impact to the organizational assets from cyber events and would prevent service outage, data loss or theft of confidential information.

A Vehicle Cyber Incident Response Management would establish Cyber Incident Response Management processes and procedures to enable the proactive cybersecurity of vehicle fleets. This means a Vehicle Incident Response Team (VCIRT) must be created by a Fleet Operator to have the ownership of developing, testing, and carrying out the Vehicle Cyber Incident Response Plan. The VCIRT needs to incorporate Automotive Cyber SMEs with the training and expertise to collect, analyze and act upon information from a vehicle cyber incident. This means that these SMEs must have a holistic skillset including In-Vehicle Security Architecture, Cyber Monitoring, Authentication & Encryption, and Incident Management best practices.

A Vehicle Cyber Incident Management process would include the following phases:

- Preparation: This includes establishment of a VCIRT and defining its responsibilities and assets in scope.
- Structure: Creation of a documented Vehicle Cyber Incidence Response Plan
- Testing: Verification & Validation of the Vehicle Cyber Incidence Response Plan
- Continuous Improvement: Ensuring that the Vehicle Cyber Incidence Response Plan is continuously enhanced

Managing Communications during a Cyber Incident

As the VCIRT "owns" the incident response, an important role it has is the communication about the incident with all other stakeholders. This includes the stakeholders within the organization, as well as external parties such as legal counsel, media, law enforcement and affected customers. Media Communications is especially important as the hacking of Connected Vehicles would become a major news story. It is, therefore, recommended that a communications specialist be part of the VCIRT and have the sole responsibility of interaction with external entities other than law enforcement. A corollary of this is that any employee of the organization who is not part of the VCIRT should not communicate about the incident with others, inside or outside the organization.

End of Vehicle Operational Lifecycle

Decommissioning is a critical part of cyber hygiene for the Vehicle Operational Lifecycle. That is why ISO 21434 assigns a complete section to this topic. Section 14 of the standard defines [9]the objective of the Decommissioning phase as

a) communicate the end of cybersecurity support; and

b) enable decommissioning of items and components with regard to cybersecurity

The key words here are "with regards to cybersecurity". What are the implications of decommissioning of the vehicle or a component in the vehicle? Is there an impact to the user? Have the cyber risks been considered?

Cyber risks of decommissioning of a vehicle or any of its components must be considered during the concept & development phases and guidelines developed to handle the decommissioning of these items. Too often, confidential information is easily obtainable from decommissioned electronics such as computers or cellphones and the same issue will translate itself into decommissioning of millions of vehicles with thousands of ECUs. Not only individual automotive manufacturers but the entire industry needs to have best practices developed around this challenge. This would guarantee that consumers feel safe & secure while getting the maximum utilization from Connected & Autonomous Vehicles (CAVs).

---

[9] ISO 21434 Road Vehicle ISO 21434 Road Vehicle Engineering

What you Learned in this Chapter

This chapter highlighted the following:

- Security by Design
- Defense in Depth
- Vulnerability Management Program
- Vehicle Cyber Monitoring
- Vehicle Cyber Incident Response Management
- End of Vehicle Operational Life Cycle

Chapter 11: The Weakest Link

Chapter Overview

This chapter enumerates the weakest link in the Cybersecurity chain i.e., human beings. The most secure system can be compromised if the human beings involved in the ecosystem of that secure system do now follow a culture of cybersecurity. A great example of this is having passwords which are not complex such as the word "password" or not applying patches based on a regular cadence. In the automotive sector, this lack of a Cyber Aware culture is a major issue as the traditional workforce was not educated to become cybersecure. This means all stakeholders need to become Cyber Aware, including the Board & CEO, Risk Management SMEs, All Employees, Suppliers, Dealers & Distributors. In the era of the Connected Vehicle, all software developers for various vehicle ECUs must also undergo secure software development training and the operators of those software must also be cognizant of the various cybersecurity challenges faced in operating a Connected Vehicle.

The Weakest Link

It is said that the most secure computer is one that is in a bunker a thousand feet underground, protected by armed guards and not connected to any internal network, let alone the Internet. The last part of this statement is especially important for securing any computer system i.e., it must be made certain that the computer system is not connected at all. To go even further, the best way for securing a computer system is to ensure it is not in any way used by human beings. This is because human beings are always the weakest link in any cybersecurity chain and can result in the breach of the most secure systems.

For the Automotive manufacturing sector that is transforming from the traditional legacy manufacturing to more tech-based "connected" vehicles, this lack of cyber awareness in its workforce is an increasingly critical challenge to resolve. This lack of cyber awareness permeates down the automotive supply chain starting from the OEM and down to the Tier 1 & Tier 2 suppliers. The lack of cyber awareness by the userbase of these vehicles of the future must also be considered. This is because such a careless approach to cybersecurity by the users would lead to actions that enable vulnerabilities in the connected vehicle. This could ultimately impact the safety of the consumers and is a major reputation risk to the OEM.

To comprehend the impact of "the weakest link", the threats posed by lack of cybersecurity awareness must be considered. There are two major distinct threats arising from the lack of cybersecurity awareness in the stakeholders of any organization. These are: Insider Threats & Social Engineering.

Insider Threats

As implied by its name, an insider threat is a security risk that originates from inside an organization i.e., from the employees of the organization itself. This is a sensitive area of discussion in any organization as employees need to be trusted and feel trusted while doing their job. However, understanding cyber risks and being a stakeholder in ensuring the safety & security of the organization is every employee's responsibility.

There are two main kinds of insider threats. The first one is the "Malicious Insider". This is the individual who might have a grudge against the company or uses his access to sensitive information to enrich himself or herself. The second kind of insider threat is the "Careless insider". This is the most common type of insider threat and arises from an employee who unknowingly enables a vulnerability into the organization's ecosystem. Ransomware is the best example of threats enabled by such "Careless Insiders" as they would unknowingly click on an insecure link and thus, infect the system with malware.

The challenge is that most cybersecurity controls are meant to stop external threats. In addition, there is a level of trust towards internal employees which most organizations are not proficient at detecting an internal threat originating from within the organization.

Social Engineering

Social Engineering is basically defined as the successful or unsuccessful attempts to influence a person(s) into either revealing information or acting in a manner that would result in unauthorized access to, unauthorized use of, or unauthorized disclosure of an information system (can be a CAV), a network, or data. It directly targets the weakest link in the security chain i.e., Human Beings and employs computers & technology merely as tools to complete the breach. Ultimately, it is the human beings themselves who are using, configuring, installing, implementing, and abusing these tools and if they themselves do not have a cybersecure mindset, then the best technology cannot secure the organization.

A great example of Social Engineering is the simple art of calling a human being directly on the phone and just asking for his or her password. This is done in creative ways, for example, mentioning that the person on the phone is from the Company Helpdesk and need to reset the users accounts. Another great example is to email an individual in the finance department and impersonate the CEO or CFO to get critical financial information about the company. This is a common way to initiate a Ransomware attack and has been successful in the automotive manufacturing sector.

Why is Social Engineering successful? The simple answer is because of human nature. Human's, by nature, are trusting of others. This is especially true if the person at the other end knows how to project an air of confidence and normality around the conversation. Today's business environment also greatly facilitates in enabling these kinds of attacks. In today's connected world, we frequently interact with individuals whom we do not know. An individual at an automotive manufacturer frequently deals with suppliers in other countries she has never met or even seen before. This is even applicable at the individual level who trusts the voice at the other end of the Telebanking service when he calls his bank to do a financial transaction.

Social Engineering attacks mostly have a defined cadence to them. The first phase is "Intelligence Gathering" in which an organization is selected for a social engineering attack and information gathered about its employees. The next phase of the attack would include "Target Selection" i.e., selecting the individual who would be the recipient of the social engineering attack. For critical infrastructure such as automotive manufacturers, this could be someone senior such as C-level individuals or it could be an individual working in a critical capacity such as the Manager responsible for new product design.

The final phase is the social engineering attack itself. This can be of three types. The first type is an "Ego Attacks" where getting the desired information could be done by manipulating the ego of a senior decision maker. The second type of attack is a "Sympathy Attack". This can be used against a person who has been passed over for promotion in a company and his resentment to the organization can be manipulated to divulge critical information. The final type of attack is the "Intimidation Attack". This, of course, is based on enabling fear in the selected individual to make him disclose the coveted information about his employer.

Changing the Culture: Invest in the People

The key to securing the weakest link is to invest in the people i.e., the workforce of the automotive manufacturing and fleet operations sector. This will become more and more critical as Connected & Autonomous Vehicles (CAVs) become more prevalent on our roads. To counter all the challenges thrown up by malicious attacks of the future, the Automotive Sector needs to develop a culture of Cyber awareness at all levels. This can only be done by educating all stakeholders in the Automotive sector and making its workforce aware of the Cyber threats and its impact to the organization.

Below are some reasons for the increase in threats to all stakeholders in the automotive sector:

- There is a greater reliance for large automotive manufacturers on *cloud infrastructure*. This means that businesses have millions of client records stored in the cloud. These form worthwhile targets for hackers looking to make a profit from selling such client records on the dark web.
- Growing sophistication of cybersecurity attackers who have incorporated *new technologies like Artificial Intelligence (AI)* to automate their attacks means that only businesses that have heavily invested in cybersecurity have adequate protection.
- Growth of apps used in the automotive sector which utilize big data, social media and web apps means that there is *an increased attack surface*

174

available to hackers to target businesses who utilize these technologies in their products.

- OEMs partnership with Tier 1 & Tier 2 Suppliers means that hackers will often **attack the Tier 1 & Tier 2 suppliers** to gain access to the networks of the OEM itself.
- There are many Start-ups in the Connected & Autonomous Vehicles (CAVs) space which do not have **adequate understanding or protection for cyber threats** to their and their client's data.

To secure against the above threats to the automotive sector and to ensure that there is a change in culture, cybersecurity training & education needs to be done at all levels of the organization and external stakeholders. Thus, a holistic Cyber Education Program needs to be developed and delivered that should include elements for all the below:

- Cyber Education for the Board & C-Suite
- Secure Software Development Education & Training
- Cyber training for Automotive Cybersecurity SMEs
- Cybersecurity Awareness for all employees of the organization
- Cyber Mentoring of the Suppliers
- Cyber Mentoring of Dealers & Distributors
- Cyber Awareness for Vehicle Operators

Cyber Education for the Board & C-Suite

One of the key factors in guaranteeing the establishment of successful Cyber strategies for an organization is to ensure that its Board of Directors & C-Suite are aware of the risks from Cyber threats. This is especially important for an automotive organization as the senior leaders play a key and long-term role in enabling this change in mindset as the industry undergoes a transformation in Mobility. C-level executives must understand the current threats and the measures that must be taken to enhance the Cyber Profile of their organization. This would also help the Board of Directors & C-level executives recognize the criticality of Cyber Governance for the smooth operations and business success of their organizations. Senior executives should be aware of the maturity level of Cyber Governance in their organization and ensure proper support, funding and resources are available to the Cyber Risk Management team to enable a long-term and holistic Cyber Governance Program in the organization.

Secure Software Development Education

With millions of lines of code in any connected vehicle, cybersecurity is a huge concern in software development for the automotive supply chain. Every bit of code can be leveraged to mount a malicious attack on the vehicle and the supply chain itself. Therefore, it is critical for OEMs and suppliers to develop holistic and detailed training programs to enable Secure Software Development.

Any Secure Software Development Education Program must educate software developers to create and maintain secure systems. This means that any such system is much harder to attack and successfully exploit to get valuable data. Such a training program must also aim to teach developers on how to reduce the impact of a successful attack by ensuring that any underlying vulnerabilities are rapidly repaired. For the automotive supply chain, it must also cover areas such as ensuring Over-the-Air (OTA) patch updates are secure. Great emphasis has been placed on the Concept & Development phase in the ISO 21434[1] standard and these should also be covered in any Secure Software Development Education Program for the automotive sector.

Cyber Training for the Automotive Cybersecurity SME

The best way to protect the automotive manufacturer from Cyber Threats is to build a specialised workforce that has the knowledge and skillset to protect the organization from myriad cyber threats. This means that the organization should define the roles & responsibilities for the Cybersecurity & Cyber Risk Management teams and hire & train qualified individuals for these roles. Although, there are many Cybersecurity trainings and certifications, these are all focused on the traditional IT security component of Cyber. There is a need for the automotive sector to develop automotive cybersecurity trainings & certifications focused on Automotive Cyber SMEs. This would enable the automotive organization to build a workforce that is resilient and futureproofs the organization from disruption & business losses from cyber risks.

---

[1] ISO 21434 Road Vehicle ISO 21434 Road Vehicle Engineering

Cybersecurity Awareness for ALL Employees of the Organization

Cybersecurity is everyone's responsibility, and a continuous & ongoing Cyber Security Awareness program ensures that everyone is aware of the risks that are faced by the organization. It is imperative for an automotive organization to ensure that all its employees undergo continuous Cyber Security Awareness training for all its employees. This is the key to changing the culture of Cyber Security in the organization and would ensure that relatively low-level cyber attacks such as ransomware & phishing attacks are easily protected against. This would facilitate the main aim of any Cyber Governance Program i.e., to ensure that any attacker must expend more time & effort to get access to its data crown jewels.

A good Cybersecurity Awareness Program should develop Policies, Awareness and Education that protects against both Insider Threats & Social Engineering. It should incentivize Cyber Hygiene by recognizing good "Catches" and test the preparedness of the Incident Response Team. It should also be testing the organizations readiness by validating whether "Targeted Groups" receive immediate notification of being possible targets of malicious attacks.

A good Cybersecurity awareness program uses many tools to disseminate information about various cybersecurity initiatives in the organization. These might include Posters, Webinar & Podcasts, Classroom Instruction, Computer-based Delivery, Brochures, Special Topic bulletins, Security banners and email notices. The key is to ensure engagement of the employees and validate their learning by measuring key metrics such as successful phishing attacks. Such activities shall ensure that there is a sea of change in the cyber culture of the organization.

Cyber Mentoring of the Suppliers

In the connected world of SmartCities & Industry4.0, it is not sufficient for a manufacturing organization to just inform its suppliers that they should meet certain cybersecurity requirements or regulations. OEMs & large Tier 1 suppliers should view their downstream suppliers as partners in enabling an eco-system which is cybersecure. This means that automotive manufacturers should educate their supply chain and work with them to ensure that there is strong collaboration between the various entities to meet any cyber challenge.

Cyber Mentoring of Dealers & Distributors

Like its suppliers, an automotive organization must work with their dealers & distributors to ensure that there is a change in culture and that these partners also become Cyber Aware. Dealers & Distributors will be the frontline soldiers in the upcoming challenge to secure Connected & Autonomous Vehicles (CAVs). Whether it is reporting of cyber incidents or managing recalls due to cyber incidents, these frontline workers need to be educated in much the same manner as they are educated about issues related to engines, brakes, powertrain, and other vehicle problems. This means that OEMS need to have a Cyber Education Program in place for its Dealers & Distributors and this program needs to be continuously monitored for Key Performance Indicators.

Cyber Awareness for Vehicle Operators

The final bit of stakeholders that need to be educated about Cyber threats to the Connected & Autonomous Vehicles (CAVs) are the owner/operators of the vehicles themselves. These might be individual car owners, or these might be huge fleet operators. However, either of these groups of people would bear the ultimate impact of any cyber hack on a connected vehicle. Some of the scenarios envisioned by Cyber SMEs for these scenarios are frightening.

Scenario 1 Ransomware 2.0: The operators of the connected vehicle might inadvertently allow ransomware by clicking on phishing emails or an app link placed in the CAV dashboard. This would lock the doors of the vehicle and the vehicle would remain being driven on the roads until and unless the passenger pays the ransomware immediately.

Scenario 2 CAV Assassin 1.0: CAV Passengers might also assist tracking of their movements by malicious individuals. This would be enabled by the passengers not following cybersecurity best practices and activating vulnerabilities which would provide such tracking data. The hacker can then use this information to direct the self-driving vehicle towards a targeted individual and the OEM could ultimately be held responsible for any crime committed using the CAV.

It is to protect against scenarios like the above, that all the stakeholders involved in the automotive ecosystem but especially OEMs need to develop cybersecurity best practices for vehicle operators and then educate & empower these individuals to apply these best practices to protect themselves from cyber incidents.

What you Learned in this Chapter

This chapter focused on enumerating the threat posed by human beings to CAV eco-system. It specifically explored the following:

- The Weakest Link
    - Insider Attacks
    - External Attacks
- Changing the Culture
    - Cyber Education for the Board & C-Suite
    - Secure Software Development Education & Training
    - Cyber training for Automotive Cybersecurity SMEs
    - Cybersecurity Awareness for all employees of the organization
    - Cyber Mentoring of the Suppliers
    - Cyber Mentoring of Dealers & Distributors
    - Cyber Awareness for Vehicle Owners/Operators

Appendices

Appendix A: Cyber Governance Policies, Procedures & Standards

Appendix   B: ISO   21434   Self-Assessment   Questionnaire

Appendix A: Vehicle Cyber Governance Policies, Procedures & Standards

Vehicle Cyber Governance Policies (Sample)

The below list is a sample of traditional IT-focused Enterprise Security Policy that would also be applicable in the CAV world:

- Access Control Policy
- Enterprise Wi-Fi Policy
- Information Security Policy
- Network Security Policy
- Payment Card Security Policy
- Public Facing Web & Application Security Policy
- Secure Software Design Policy
- Security Logging & Monitoring Policy
- Server Configuration Security Policy
- Supplier Cybersecurity Policy
- Threat Identification & Management Policy
- Vulnerability Management Policy
- Workstation Configuration Security Policy
- Security Awareness & Education Policy
- Threat, Risk & Vulnerability Assessment (TRVA) Policy
- Cryptography & Digital Certificates Policy
- Business Continuity Planning Policy
- Anti-Malware Policy
- Data Center Protection Policy
- Physical Security Policy
- Cyber Incident Management Policy
- Information Asset Management Security Policy
- Mobile Device (BYOD) Security Policy

- Password Security Policy
- Patch Management Security Policy
- Remote Work Security Policy

For the Automotive Sector, besides the applicability of the above-mentioned policies, the following additional cybersecurity policies should be considered:

- OT Cybersecurity Policy
- Fleet Cybersecurity Policy
- In-Vehicle Security Policy
- Vehicle Cyber Incident Management Policy
- V2X Security Policy
- CAV Threat Analysis & Risk Assessment (TARA) Policy
- CAV Third-Party Apps Cyber Risk Management Policy

Vehicle Cyber Governance Procedures (Sample)
Below are some of the sample traditional IT Cyber Governance procedures:

- Change Management Procedure
- Employee Departure Procedure
- Device Management Review Procedure
- Cyber Incident Management Procedure
- Incident Management Procedure
- Linux Patch Management Procedure
- New Hire Account Management Procedure
- System Account Procedure
- Vulnerability Management Procedure

For the Automotive Sector, besides the applicability of the above-mentioned procedures, the following additional cybersecurity procedure should be considered:

- CAV Cyber Breach Notification Procedure
- ECU Threat Modeling Procedure

182

- ECU Risk Assessment Procedure
- ECU Security Configuration Procedure

Vehilce Cyber Governance Standards/Benchmarks (Sample)

Below are some of the traditional IT Cyber Governance standards/benchmarks:

- Cryptography & Digital Certificates Standard
- Server Configuration Standard
- Workstation Configuration Standard

For the Automotive Sector, besides the applicability of the above-mentioned standards, the following additional cybersecurity standards should be considered:

- In-Vehicle Security Configuration Standard
- V2X Security Standard
- OT Security Standard
- ECU Cybersecurity Standard

Appendix B: ISO 21434 Self-Assessment Questionnaire

Section 5: Organizational Cybersecurity Management

This section outlines the steps needed to ensure overall Cybersecurity Governance in the Organization.   It also enumerates the steps that are needed to change the overall Organization Culture to make it more Cyber aware.

Work Product	Questions to be asked	Requirement	Compliant / Non-Compliant
[WP-05-01] *Cybersecurity policy, rules, and processes*	Does the Organization have a cybersecurity policy that acknowledges the road vehicles cyber risks and the Executive Management's commitment to protecting the Organization assets from these risks to the Organization?	**RQ-05-01**	
	Does the Organization have a formal Cyber Security Management System (CSMS) in place?	**RQ-05-02**	
	Does the Cyber Security Management System (CSMS) enumerate all relevant Cyber rules & processes, Cyber responsibilities & the resources needed to enable Cyber in the Organization?	**RQ-05-02**	
	Does the Cyber Security Management System (CSMS) assign roles and responsibilities for protecting	**RQ-05-03**	

	data assets in the organization?		
	Have appropriate resources been assigned to ensure the Cybersecurity responsibilities are effectively carried out?	**RQ-05-04**	
	Does the Cyber Security Management System (CSMS) specify the stakeholders involved in protecting road vehicles data assets in the Organization? And have their Cyber responsibilities been effectively communicated?	**RQ-05-05**	
[WP-05-02] Evidence of competence management, awareness management	Is there a formal program in place to change the Organization culture to have a more cyber-oriented mindset?	**RQ-05-06**	
	Are competent and knowledgeable personnel responsible for Cyber Activities?	**RQ-05-07**	
[WP-05-03] Evidence of the organization's management systems	Has the Organization instituted and maintains a continuous improvement process for cybersecurity?	**RQ-05-08**	
	Does the Organization have clear guidelines on sharing of Data Assets?	**RQ-05-09** **RC-05-10**	
	Does the Organization have a Quality Management System (QMS)?	**RQ-05-11**	
	Does the Organization have a Configuration Management System (CMS)?	**RQ-05-12**	

	Does the Organization have a Cyber Security Management System (CSMS)?	**RC-05-13**	
[WP-05-04] Evidence of Tool Management	Does the Organization have a list of Tools that may impact Cybersecurity?	**RQ-05-14**	
	Are these Tools managed to support remedial actions for cybersecurity incidents?	**RC-05-15**	
[WP-05-05] Organizational cybersecurity audit report	Are work products being maintained as per the CSMS?	**RC-05-16**	
	Are cybersecurity audits being independently performed on periodic basis?	**RQ-05-17**	

Section 6: Project dependent Cybersecurity Management

This section details the requirements to manage projects related to Cybersecurity in the Organization.

Work Product	Questions to be asked	Requirement	Compliant/ Non- Compliant
[WP-06-01] Cybersecurity plan	Are responsibilities for defining project cybersecurity activities assigned and communicated in accordance with CSMS of the organization?	**RQ-06-01**	
	Has there been a formal gap analysis carried out to define the activities detailed in the Project Cybersecurity Plan for the specific component or item?	**RQ-06-02**	
	Does the Organization have a formal documented Project Cybersecurity Plan in place for the project? Does the Project Cybersecurity Plan specify the activities required, the roles & responsibilities, the dependencies, the desired outcomes, and the Work Products needed for relevant in scope assets?	**RQ-06-03**	
	Is there specific owner defined for developing the project cybersecurity plan? Are project cybersecurity activities being tracked?	**RQ-06-04**	
	Is the Project Cybersecurity	**RQ-06-05**	

	Plan referenced in the overall Project Plan?		
	Does the Project Cybersecurity Plan specify the activities that are required for cybersecurity during the concept and product development phases?	**RQ-06-06**	
	Is the Project Cybersecurity Plan continuously updated and validated?	**RQ-06-07**	
	Is the determination of Threat Values for various risk levels determined as part of the Project Cybersecurity Plan?	**PM-06-08**	
	Are the Work Products identified in the Project Cybersecurity Plan updated and maintained for accuracy?	**RQ-06-09**	
	Is there a Project Cybersecurity Plan for both the customer and supplier?	**RQ-06-10**	
	Are both the Project Cybersecurity Plan and related work products subject to configuration management, change management, requirements management, and documentation management of the organization?	**RQ-06-11** **RQ-06-12**	
	If any cybersecurity activity is tailored, rational for why that tailored activity is sufficient should be provided?	**PM-06-13** **PM-06-14**	
	Are reuse analysis carried out for any item or components that have been modified?	**RQ-06-15** **RQ-06-16** **RQ-06-17**	
	Are cybersecurity	**RQ-06-18**	

	assumptions for out-of-context components clearly articulated?	RQ-06-19 RQ-06-20	
	Is cybersecurity documentation gathered for off-the-shelf components?	RQ-06-21 RQ-06-22	
[WP-06-02] Cybersecurity case	Has a formal Cybersecurity Case been developed based on the Project Cybersecurity Plan to ensure alignment with the Organization's Business Vision & Mission?	RQ-06-23	
[WP-06-03] Cybersecurity assessment report	Has a Cybersecurity assessment been carried out for the item or component under development?	RQ-06-24	
	Has the Cybersecurity assessment been carried out by an independent individual or entity?	RQ-06-25 RQ-06-26 RQ-06-27 RQ-06-28	
	Does the Cybersecurity assessment provide confidence that the achieved degree of cybersecurity of the item or component is sufficient?	PM-06-29	
	Has the scope of the Cybersecurity assessment been adequately defined?	RQ-06-30	
	Has a formal cybersecurity assessment report been issued? Does the cybersecurity assessment report include a recommendation for the acceptance, conditional acceptance, or rejection of	RQ-06-31 RQ-06-32	

	the achieved degree of cybersecurity of the item or component?		
[WP-06-04] Release for post-development report	Are the conditions met for release of the report including the Cybersecurity case, the cybersecurity assessment report, and the cybersecurity requirement for post-development? Does the report ensure that the cybersecurity requirements for the project are met?	**RQ-06-33** **RQ-06-34**	

Distributed Cybersecurity Activities

Section 7: Distributed Cybersecurity Activities

This section outlines the Cybersecurity requirements for the supply chain.

Work Product	Questions to be asked	Requirement	Compliant / Non-Compliant
[WP-07-01] Cybersecurity interface agreement	Has the Organization assessed the Cybersecurity capabilities of all its suppliers by analyzing the Supplier record of capability?	**RQ-07-01** **RC-07-02**	
	Do all Request for Quotations include the expectation of Cyber responsibilities from the suppliers?	**RQ-07-03**	
	Are Cybersecurity Interface Agreements executed with all suppliers of the Organization? Does the Cybersecurity Interface Agreements include the roles and responsibilities of both the Customer and the Supplier through a Responsibility Assignment Matrix?	**RQ-07-04** **RC-07-05** **RC-07-08**	
	Are the responsibilities regarding Vulnerability Management clearly outlined in the Cybersecurity Interface Agreement?	**RQ-07-06**	
	Is a notification process defined in the Cybersecurity Interface Agreement to communicate unforeseen cybersecurity issues?	**RQ-07-07**	

Continual Cybersecurity Activities

Section 8: Continual Cybersecurity Activities

Continual Cybersecurity activities are those activities which are required to be performed during all phases of the item or components lifespan and are not specific to a defined project.

Work Product	Questions to be asked	Requirement	Compliant / Non-Compliant
[WP-08-01] Sources for cybersecurity information	Have all internal sources been identified including information received from the field?	**RQ-08-01**	
	Have all external sources of Cybersecurity been identified, including organization's customers & suppliers?	**RQ-08-01**	
	Are all internal & external sources of Cybersecurity information monitored for Cybersecurity events?	**RQ-08-01**	
[WP-08-02] Triggers	Is there a formal methodology in place to define and maintain the triggers?	**RQ-08-02**	
[WP-08-03] Cybersecurity events and triage of cybersecurity information	Is Cybersecurity information collected from all relevant sources?	**RQ-08-03**	
	Is there a formal methodology defined to ensure triage of Cybersecurity Information?	**RQ-08-03**	
	Have criteria for triage been	**RQ-08-03**	

	defined that can be used to distinguish trigger thresholds?		
	Are results from the triage of Cybersecurity information readily available for further action?	**RQ-08-03**	
[WP-08-04] Weakness from cybersecurity events	Has the Cybersecurity event been evaluated to identify weakness in an item or/component?	**RQ-08-04**	
	Has a Risk Treatment decision been taken with respect to that specific item or component?	**RQ-08-04**	
[WP-08-05] Vulnerability analysis	Are cybersecurity vulnerabilities identified using vulnerability analysis?	**RQ-08-05** **RQ-08-06**	
[WP-08-06] Vulnerability Management	Is a Vulnerability Management Program in place to mitigate & manage identified vulnerabilities?	**RQ-08-07**	
	Does the Vulnerability Management Program manage identified vulnerabilities based on identified risk associated with these vulnerabilities?	**RQ-08-08**	

Concept & Product Development Phases

Section 9: Concept Phase

This section defines the item, the need to carry out a TARA to develop a RTP, the definition of Cybersecurity Goals and Cybersecurity Claims for each specified item.

Work Product	Questions to be asked	Requirement	Compliant / Non-Compliant
[WP-09-01] Item definition	Has item definition been carried out to ensure a formal implementation of Cyber best practices?	**RQ-09-01**	
	Does the item definition include item boundary, function, and architecture?	**RQ-09-01**	
	Does the item definition include operational environment of the item with respect to cybersecurity?	**RQ-09-02**	
	Does the item definition include constraints & compliance needs of the specific item?	**RQ-09-02**	
	Does the item definition enumerate any assumptions made during the item definition process?	**RQ-09-02**	
[WP-09-02] Threat analysis and risk assessment	Has a TARA been conducted for the defined item that includes all assets encompassed by the specified item?	**RQ-09-03**	

	Is there a Risk Treatment Plan that enumerates Risk treatment decisions for identified Threat Scenarios and their associated risks?	**RQ-09-04**	
[WP-09-03] Cybersecurity goals	Has Cybersecurity goals, such as CAL, been established based on the Risk Treatment Plan for each specified item?	**RQ-09-05**	
[WP-09-04] Cybersecurity claims	Have any Cybersecurity claims been stated for the operational environment that leads to reduction of risk for a Threat Scenario?	**RQ-09-06**	
	Have any Cybersecurity claims been stated for any risk treatment options that leads to sharing or transferring risk?	**RQ-09-06**	
[WP-09-05] Verification report for cybersecurity goals	Is the process to determine Cybersecurity Goals & Cybersecurity Claims verified through a well documented report?	**RQ-09-07**	
[WP-09-06] Cybersecurity concept	Has the Cybersecurity concept been documented that specifies the cybersecurity requirements needed to meet the Cybersecurity Goals of the specified item?	**RQ-09-08** **RQ-09-09** **RQ-09-10**	
[WP-09-07] Verification report of cybersecurity concept	Has the Cybersecurity concept been verified through a formal report?	**RQ-09-11**	

Concept & Product Development Phases

Section 10: Product Development

This section describes the cybersecurity requirements of the architecture design as well as the integration and verification activities need to ensure those cybersecurity requirements are met.

Work Product	Questions to be asked	Requirement	Compliant / Non-Compliant
[WP-10-01] Cybersecurity specification	Have the Cybersecurity requirements been defined for Product development based on cybersecurity requirements allocated at a higher level?	**RQ-10-01**	
	Have the Cybersecurity requirements been defined for Product development based on architecture design from a higher level?	**RQ-10-01**	
	Do the Cybersecurity requirements include any applicable cybersecurity controls?	**RQ-10-01**	
	Has the architecture designed been refined to ensure applicability of various Cybersecurity requirements?	**RQ-10-01**	
	Have the interfaces between the components of the refined architecture design that are applicable to meet cybersecurity requirements been identified?	**RQ-10-02**	
[WP-10-02]	Have the cybersecurity	**RQ-10-03**	

Cybersecurity requirements for post-development	implications for the post-development phase been considered while enumerating the cybersecurity requirements?		
	Have specific cybersecurity requirements been formally documented to ensure cybersecurity in the post-developmental phase?	**RQ-10-03**	
[WP-10-03] Documentation of the modelling design or programming languages and coding guidelines, if appliable	Has a Criteria for suitable design, modelling and programming languages for cybersecurity been established?	**RQ-10-04** **RQ-10-05**	
[WP-10-04] Verification report for the cybersecurity specification	Have the refined cybersecurity requirements and the refined architecture design been verified through a formal documented report?	**RC-10-06** **RQ-10-07**	
[WP-10-05] Weaknesses found during product development	Has an evaluation been carried out to identify weaknesses in the refined architecture design and the cybersecurity requirements?	**RQ-10-08**	
[WP-10-06] Integration and verification specification	Have Integration & Verification specification been defined for the development phase?	**RQ-10-09** **RQ-10-10**	
[WP-10-07]	Have Verification testing	**RQ-10-11**	

Integration and verification reports	been performed to ensure compliance with the refined Cybersecurity requirements?	**RQ-10-12** **RQ-10-13**	
	Are the Integration & Verification activities outlined in a formal documented report?	**RQ-10-11** **RQ-10-12** **RQ-10-13**	

Concept & Product Development Phases

Section 11: Cybersecurity Validation

This section outlines the Cybersecurity Validation of the Item at Vehicle Level

Work Product	Questions to be asked	Requirement	Compliant / Non-Compliant
[WP-11-01] Validation report	Have Validation activities been carried out at the Vehicle level to ensure adequacy of cybersecurity goals and validity of cybersecurity claims?	**RQ-11-01**	
	Has a Validation specification been developed to define these Validation activities?	**RQ-11-01**	
	Has Penetration Testing been performed to validate the Cybersecurity Goals?	**RQ-11-01**	
	Has a Validation report been generated which details the risk identified and their acceptance rationale for a specific item during the Concept & Product Development phases?	**RQ-11-02**	

Post-Development Phases

Section 12: Production

This section ensures that the Cybersecurity requirements are applied during the production process and that no further vulnerabilities creep into the item or component during this process.

Work Product	Questions to be asked	Requirement	Compliant / Non- Compliant
[WP-12-01] Production control plan	Has a Production Control Plan been created to apply the Cybersecurity requirements during the production process?	**RQ-12-01**	
	Does the Production Control Plan include the sequence of steps that apply the cybersecurity requirements for post-development?	**RQ-12-02**	
	Does the Production Control Plan include production tools and equipment relevant to implementing the cybersecurity requirements for post-development?	**RQ-12-02**	
	Does the Production Control Plan include details on how to protect the item or component for unauthorized alteration during production?	**RQ-12-02**	
	Does the Production Control Plan include activities to validate that Cybersecurity requirements are met during the production process?	**RQ-12-02**	
	Is the Production Control Plan implemented?	**RQ-12-03**	

Post-Development Phases

Section 13: Operations & Maintenance

This section describes the activities required for enabling relevant Cyber activities during the Operations & Maintenance of the Road Vehicle.

Work Product	Questions to be asked	Requirement	Compliant / Non-Compliant
[WP-13-01] Cybersecurity incident response plan	Is there a Cybersecurity Incident Plan in place to handle Cybersecurity incidents?	RQ-13-01	
	Does the Cybersecurity Incident Plan define the remediation actions required to handle the Cybersecurity incident?	RQ-13-01	
	Does the Cybersecurity Incident Plan include an effective communication plan during the Cybersecurity incident?	RQ-13-01	
	Does the Cybersecurity Incident Plan define the roles & responsibilities to remediate the incident?	RQ-13-01	
	Does the Cybersecurity Incident Plan include a procedure for recording new cybersecurity information relevant to the cybersecurity incident?	RQ-13-01	
	Does the Cybersecurity Incident Plan include a method for determining progress for remediating the Cybersecurity incident?	RQ-13-01	

	Does the Cybersecurity Incident Plan include criteria for closing the Cybersecurity Incident and actions to be undertaken at this stage?	RQ-13-01	
	Is all information relevant to a specific Cybersecurity incident gathered in a formal manner to ensure appropriate Cybersecurity incident response?	RQ-13-01	
	Is the Cybersecurity Incident Plan implemented?	RQ-13-02	
Updates	Are updates developed based on the Cybersecurity Goals of the item or components?	RQ-13-03	
	Are Cybersecurity implications of recovery options considered while carrying out updates?	RQ-13-03	

Post-Development Phases

Section 14: End of cybersecurity support and decommissioning

This section outlines the Cybersecurity implications of Decommissioning activities.

Work Product	Questions to be asked	Compliant	Non-Compliant
[WP-14-01] Procedures to communicate the end of cybersecurity support	Are procedures in place to communicate the end of cybersecurity support for a particular item?	**WP-14-01**	
Decommissioning	Are Cybersecurity implications of Decommissioning activities considered at the time of Decommissioning? These should be part of the cybersecurity requirements for post-development [WP-10-02].	**RQ-14-02**	
	Is an item or component decommissioned in a secure manner?	**RQ-14-02**	

Section 15: Threat analysis & risk assessment methods

This section defines the Threat Analysis & Risk Assessment (TARA) to be used to ensure a formal Risk Treatment Plan is developed to address the risks associated with each Threat Scenario.

Work Product	Questions to be asked	Requirement	Compliant / Non-Compliant
[WP-15-01] Damage scenarios	Is there a process in place to identify "Damage Scenarios" for assets involved in functioning of road vehicles?	RQ-15-01	
[WP-15-02] Assets with cybersecurity properties	Is there an asset inventory in place that identified all assets related to road vehicles in the Organization?	RQ-15-02	
	Is it ensured that assets with cybersecurity properties whose compromise leads to damage scenarios are detailed in the asset inventory?	RQ-15-02	
[WP-15-03] Threat scenarios	Are Threat scenarios applicable to the targeted assets identified? Does these include the compromised cybersecurity property of the asset as well as the cause of compromise of the cybersecurity property?	RQ-15-03	
[WP-15-04] Impact rating, including the associated impact	Are potential adverse consequences to the road users in terms of safety, financial, operational, and privacy (S,F,O,P) assessed for	RQ-15-04	

categories of the damage scenarios	various Damage Scenarios?		
	Are Impact Ratings (severe, major, moderate, negligible) identified for the various Damage Scenarios for each independent category i.e. safety, financial, operational, and privacy (S,F,O,P)?	RQ-15-05	
	Are Safety related impact ratings shall be derived from ISO 26262-3:2018, 6.4.3?	RQ-15-06	
[WP-15-05] Attack paths	Are Attack Paths analysed for each Threat Scenarios?	RQ-15-08	
	Are attack paths associated with the threat scenarios that can be realized by the attack path	RQ-15-09	
[WP-15-06] Attack feasibility rating	For each Attack Path, has the Attack feasibility rating (High, Medium, Low, Very Low) been determined?	RQ-15-10	
	Has an attack feasibility method been determined?	RC-15-11 RC-15-12 RC-15-13 RC-15-14	
[WP-15-07] Risk values	Has the Risk Value for each Threat Scenario been determined?	RQ-15-15	
	Have risk matrices or risk values been finalized so that the risk value of a threat scenario shall be a value between (and including) 1 and	RQ-15-16	

	5, where a value of 1 represents minimal risk		
[WP-15-08] Risk treatment decision per threat scenario	Has Risk Treatment Options been analysed and determiend for each Threat Scenario based on its impact categories, attack paths and Risk Value?	**RQ-15-17**	
	Are these Risk Treatment Options formally documented in a Risk Treatment Plan?	**RQ-15-17**	

This page is intentionally left blank

Manufactured by Amazon.ca
Bolton, ON

33703883R00122